THE COMMONWEALTH AND INTERNATIONAL LIBRARY
Joint Chairmen of the Honorary Editorial Advisory Board
SIR ROBERT ROBINSON, O.M., F.R.S., LONDON
DEAN ATHELSTAN SPILHAUS, MINNESOTA
Publisher: ROBERT MAXWELL, M.C., M.P.

SELECTED READINGS IN PHYSICS
General Editor: D. TER HAAR

EARLY SOLAR PHYSICS

EARLY SOLAR PHYSICS

BY

A. J. MEADOWS

Senior Lecturer in Astronomy
University of Leicester

1966

PERGAMON PRESS

OXFORD · LONDON · EDINBURGH · NEW YORK
TORONTO · SYDNEY · PARIS · BRAUNSCHWEIG

PERGAMON PRESS LTD.,
Headington Hill Hall, Oxford
4 & 5 Fitzroy Square, London W.1

PERGAMON PRESS (SCOTLAND) LTD.,
2 & 3 Teviot Place, Edinburgh 1

PERGAMON PRESS INC.,
Maxwell House, Fairview Park, Elmsford, New York 10523

PERGAMON OF CANADA LTD.,
207 Queen's Quay West, Toronto 1

PERGAMON PRESS (AUST.) PTY. LTD.,
19a Boundary Street, Rushcutters Bay, N.S.W. 2011, Australia

PERGAMON PRESS S.A.R.L.,
24 rue des Écoles, Paris 5e

VIEWEG & SOHN GMBH,
Burgplatz 1, Braunschweig

First edition 1970

Library of Congress Catalog Card No. 74–103021

Printed in Great Britain by Thomas Nelson (Printers) Ltd, London and Edinburgh

08 006654 2 (flexicover)
08 006653 4 (hard cover)

Contents

v

vi   CONTENTS

Preface

THE growth of solar physics is a vast and intricate subject full of twists and turns. The description of it provided in this book is no more than a brief, selective introduction to the early period. The emphasis here is more especially on the development of solar spectroscopy, and on the relationships which were discovered between the various layers of the solar atmosphere and between the different forms of solar activity. In terms of time, the main concentration is on the years from 1850 to 1900. This is, of course, an arbitrary choice, but it represents, nevertheless, a period during which knowledge of the Sun progressed from the extremely rudimentary to the recognizably modern. The fact that a new type of astronomy was being created received full contemporary recognition. S. P. Langley, writing in 1888, noted that:

"Within a comparatively few years, a new branch of astronomy has arisen. . . . Its study of the Sun, beginning with its external features (and full of novelty and interest, even as regards those), led to the further enquiry as to what it was made of, and then to finding the unexpected relations which it bore to the earth. . . . This new branch of inquiry is sometimes called Celestial Physics, sometimes Solar Physics and is sometimes . . . referred to as the New Astronomy."

To provide some insight into the changes and developments which a single individual might feel forced to make to his views of the Sun during this period, the work of Sir Norman Lockyer is described separately in rather greater detail.

This main body of the commentary is contained in Chapter II. Chapter I is a short introduction, setting the scene for the following period of rapid growth. Chapter III is by way of an epilogue. If it is necessary to distinguish between early and modern solar physics, the division may, perhaps, be set most reasonably in 1913, when Bohr proposed the first workable theory of atomic spectra. The third chapter selects two advances during the first decade of the present century—one theoretical and one observational—to show how solar physics was ready to move ahead into the modern era as soon as this vital development in physics occurred.

I am much indebted to the following organizations for permission to reprint articles in Part 2 of this book: the University of Chicago Press for the chapter by G. E. Hale from *Stellar Evolution* and for the article by W. W. Campbell in the *Astrophysical Journal* (1899); the Smithsonian Institution for the article by G. E. Hale in the *Smithsonian Report* (1913); the Royal Society for the article by J. Evershed in the *Proceedings of the Royal Society* (1901).

Part 1

CHAPTER I

Ideas of the Sun in the Mid-Nineteenth Century

THE initial burst of enthusiasm for solar observation, which followed the introduction of the astronomical telescope at the beginning of the seventeenth century, waned fairly quickly. The discoveries made during this early period included observation of sunspots, of faculae and of solar rotation; no further advances of importance in solar astronomy were made subsequently until the latter half of the eighteenth century. This lack of progress may be attributable in part to a certain scarcity of sunspots throughout much of this period, for it was the spots which caused much of the early interest in the Sun, and they, too, were involved in the resurgence of interest in the later eighteenth century

In 1774, Alexander Wilson, a professor at Glasgow University, published some observations which seemed to show conclusively that sunspots were neither clouds floating above the solar surface, nor layers of accumulated slag, nor volcanoes—all of which had been suggested—but were, instead, depressions below the normal level of the surface (Wilson, 1774). He obtained this result from a study of the apparent change in appearance of a spot as it crossed the solar disc.

"Astronomers will remember that a spot of an extraordinary size appeared upon the sun, in the month of November 1769. . . . on the 22nd day I had a view of the sun through an excellent Gregorian telescope. . . . I then beheld the spot which at that time was not far from the sun's western limb. . . . Next day being the 23rd, I had a curiosity to see it again. . . . I now found however a remarkable change; for the umbra, which before was equally broad all round the nucleus, appeared much contracted *on that part which lay towards the centre of the disc*. . . . I began to suspect that the central part, or nucleus of this spot, was beneath the level of the sun's spherical surface; and that the shady zone or umbra, which surrounded it, might be nothing else but the shelving sides of the luminous matter of the sun, reaching from his surface, in every direction, down to the nucleus; for, upon this supposition, I perceived, that a just account could be given of the changes, of the umbra and of the figure of the nucleus.'†

The existence of this Wilson effect was disputed by his influential French contemporary, Lalande, who preferred to believe that spots represented mountains sticking up through the liquid surface of the Sun. Wilson's idea was, however, taken up in England by Sir William Herschel (1795), who developed it into a general description of the solar constitution. He believed the bright surface of the Sun to be the upper side of a layer of luminous clouds; the spots were regions where the clouds had been temporarily dispersed by atmospheric currents, so that the dark, solid body of the true Sun was exposed. In detail, his theory became more complicated, for two cloud layers were actually envisaged: the opening in the upper layer being larger than that in the lower. In this way, the presence of a penumbra round the central umbra of a spot could be explained. Herschel speculated that, since the supposed true surface of the Sun was protected from the heating effects of its luminous envelope by the lower-lying clouds, therefore the solar surface could be inhabited. This last opinion was generally disregarded, but his overall concept of the solar constitution was widely accepted until after the middle of the nineteenth century.

". . . according to the above theory, a dark spot in the Sun is a place in its atmosphere which happens to be free from luminous decompositions; and that faculae are, on the contrary, more copious mixtures of such fluids as decompose each other. The

† Note that Wilson refers to the umbra of the spot as its nucleus, and calls the penumbra the umbra.

penumbra which attends the spots, being generally depressed more or less to about half way between the solid body of the sun and the upper part of those regions in which luminous decompositions take place, must of course be fainter than other parts.

The sun, viewed in this light, appears to be nothing else than a very eminent, large and lucid planet, evidently the first, or in strictness of speaking, the only primary one of our system; all others being truly secondary to it. Its similarity to the other globes of the solar system with regard to its solidity, its atmosphere, and its diversified surface; the rotation upon its axis, and the fall of heavy bodies, leads us on to suppose that it is most probably also inhabited like the rest of the planets, by beings whose organs are adapted to the peculiar circumstances of that vast globe."

Slight modifications were introduced into this scheme from time to time. In 1851, for example, Rev. W. R. Dawes—one of the leading British solar observers—detected small, round spots in the umbra which were even darker than their surroundings. He therefore assumed that the umbra of a spot did not, in fact, represent the true solar surface. It was, instead, an even lower cloud stratum, and only the smaller spots he had discovered provided a true glimpse of the surface (Dawes, 1852).

The apparent similarity between the solar and terrestrial atmospheres suggested a comparison between solar and terrestrial meteorology. Sir John Herschel (1847), for example, thought that solar rotation would produce a higher temperature at the equator of the Sun than at the poles. He was thus led to suppose that atmospheric zonation occurred on the Sun, as on the Earth. Sunspots were then comparable with terrestrial cyclones: they represented regions where the cloudy strata of the Sun were penetrated by downward motions of the solar atmosphere. (This contrasted with the opposing view at the time that the clearings were due to an upward motion of the atmosphere as a result of volcanic explosions on the solar surface.) The idea that sunspots could be identified with some kind of vortex motion continued to be more or less influential throughout the next hundred years.

The picture of the Sun as a relatively cool, solid body surrounded by layers of luminous clouds gradually disappeared as the full implications of new work, particularly in the field of spectroscopy, were understood. Nevertheless, it lingered on for a surprisingly long time: Sir John Herschel's standard textbook—*Outlines of Astronomy*—still reproduced Sir William Herschel's scheme at the end of the 1860's.

It was apparent from the earliest solar observations that the number of sunspots visible varied with time, but the first detailed, long-term survey of these variations was not undertaken until well into the nineteenth century. In the early 1840's, S. H. Schwabe, an amateur astronomer of Dessau, announced the results from seventeen years observations of sunspot numbers (the paper is reprinted in Part 2). He showed that the number of spots visible on the Sun's surface, instead of varying randomly as had usually been supposed, rose and fell with a period of about ten years. His initial announcement in the *Astronomische Nachrichten* aroused only slight interest, but Humboldt was struck by his results and published them in the volume of his book *Kosmos* which appeared in 1851. This work was very widely read, so that the concept of a sunspot cycle quickly became general currency. At about the same time, Lamont (1852) at the University of Munich published an analysis of observations of terrestrial magnetism made in Germany over the previous 15 years. He pointed out that the amplitude of the daily variation was itself variable, with a period of $10\frac{1}{3}$ years. He did not, however, relate this discovery to Schwabe's sunspot cycle (indeed, he later opposed the belief that any such connection existed). Very shortly afterwards, Sir Edward Sabine (1852) began an examination of magnetic observations from Canada. He concentrated on the occurrence of magnetic storms, and found that they varied in frequency and amplitude with a periodicity of about ten years. Unlike Lamont, he recognized the relationship between his results and those of Schwabe for sunspots, and pointed out further that the magnetic fluctuations he had been examining were greatest when sunspot numbers were at a maximum.

". . . new and important features have presented themselves in the comparison of the frequency and amount of the disturbances in *different years*, apparently indicating the existence of a *periodical variation*, which, either from a causal connection (meaning thereby there being possibly joint effects of a common cause), or by a singular coincidence, corresponds precisely both in period and epoch, with the variation in the frequency and magnitude of the solar spots, recently announced by M. Schwabe as the result of his systematic and long-continued observations."

Sabine's identification of the connection between the sunspot and the magnetic periods was published in mid-1852. Within the next few months the same identification was also made on the Continent by Wolf (1852) and, independently, by Gautier (1852) These publications mark the beginning of research into solar-terrestrial relations.

As we have remarked, spots were not the only solar features noted by early telescopic observers. Faculae—small bright patches, best seen near the Sun's limb—were recognized at about the same time. Similarly, it was soon observed that the Sun's surface was mottled in appearance rather than uniform. By the middle of the nineteenth century the darkening of the solar disc towards the limb was also well known, although still slightly controversial—Arago, for example, believed that genuine limb-darkening only existed to a very limited extent. This question was to some extent settled when the first good daguerreotype of the Sun was obtained by Fizeau and Foucault in 1845 (at Arago's request). This clearly showed limb-darkening, as well as the umbrae and penumbrae of some spots present (see the reproduction in G. de Vaucouleurs (1961), Plate 1).

A few measurements of the overall brightness of the Sun had been made during the eighteenth century (e.g. by Bouguer in 1725 and by Wollaston in 1799). The method used was very elementary —a visual comparison of a candle flame with the Sun in the meridian—and the results were correspondingly inaccurate. There was also considerable doubt as to the magnitude of many of the corrections (such as that for the atmospheric absorption of sunlight) which had to be made. In 1844, Foucault and Fizeau tried to obtain a more refined estimate of the solar brightness by comparing daguerreotype images of the Sun and incandescent limelight. Their results, however, proved to be no more reliable than the earlier visual estimates, which were therefore still being quoted at the mid-century.

The first quantitative estimate of the heat emitted by the Sun was made by Newton (who also reasoned that the Sun must be red-hot throughout). The first significant measurements, however, were not made until well into the nineteenth century. Then in 1837–8 Sir John Herschel (1847), during his stay at the Cape of Good Hope, carried out some observations with an actinometer. (This consisted essentially of a bowl of water which was exposed to the Sun for a short period of time, and the resulting rise in temperature noted.) About the same time, C. S. M. Pouillet in France began a series of measurements with a pyrheliometer (based on essentially the same principle as the actinometer, though considerably different in appearance). Their results for what was named the solar constant agreed quite well, but were, in fact, only slightly over half the value accepted later in the century. (For a description and comparison of Herschel's and Pouillet's work, see: Young, 1896, chapter 9.) Shortly afterwards, J. D. Forbes (1842) obtained a much higher value, but it was considered at the time to be less reliable. As in the measurements of solar brightness, one of the main experimental difficulties lay in the allowance for atmospheric absorption.

An important reason for determining the solar constant was that it could, theoretically, be used to derive a surface temperature for the Sun. The major difficulty, which plagued scientists throughout the nineteenth century, concerned the exact relationship between the temperature of a surface and the amount of radiation it emitted. Until the early part of the nineteenth century, Newton's law of cooling had been universally accepted, i.e. the radiation emitted was supposed to be directly proportional to the temperature. Then, however, the French scientists, Dulong and Petit (1817), carried out a series of experiments which showed fairly conclusively that Newton's law seriously underestimated the amount of heat emitted. They proposed instead an empirical formula which, in effect, equated an arithmetical increase in temperature to a geometrical increase in the radiation emitted. Both laws were in use throughout the nineteenth century for determining the surface temperature of the Sun, with the result that the various estimates differed enormously. The difficulty with both laws, of course, was that they had only been tested over a relatively small range of temperatures in the laboratory: it was generally realized that extrapolation to the sort of temperature found on the Sun might not be justifiable, but nothing better was available. Some of the results were obviously impossible. Thus Pouillet used his determination of the solar constant, together with Dulong and Petit's law, to determine the Sun's temperature. He arrived at a figure of 1750° C. It was pointed out that this corresponded to the melting point of platinum, yet it was known that platinum could be vaporized by a large burning glass. Hence, his result was necessarily too low.

In 1845, J. Henry at Princeton investigated the relative amount of heat coming from the centre and limb of the Sun. With his encouragement, similar observations, but in greater detail, were made by Secchi (1852) who claimed that he had detected an appreciable difference in the amount of radiation coming from the equator of the Sun and from higher latitudes. He also believed that he had found a significant difference between the northern and southern hemispheres of the Sun.

Solar eclipses have, of course, been observed accidentally from time immemorial. They were not, however, subjected to a special study until the nineteenth century. It is, perhaps, possible to attribute the interest which then appeared partly to the general revival of solar studies and partly to the relative ease of travel to eclipse sites as compared with earlier centuries. The systematic study of solar eclipses may reasonably be dated from the mid-1830's, when Francis Baily (1836) observed an annular eclipse of the Sun. He was startled by the sudden appearance of a string of bright points along the limb of the Moon as it moved across the solar disc. This phenomenon, which he attributed to the effect of sunlight shining between lunar mountains, is usually referred to as Baily's beads. Although it had been observed before, the earlier accounts had been generally forgotten and Baily's report aroused considerable interest amongst his fellow astronomers, helping to promote extensive preparations for the total eclipse of 1842. The observations of this eclipse, which was visible over southern Europe, in turn, sparked off the numerous expeditions to subsequent eclipses which were an important part of nineteenth-century astronomy.

The chromosphere, the corona and prominences had all been noted prior to 1842, but the earlier observations do not seem to have aroused any very great curiosity (perhaps because there was a general tendency to consider them all as optical illusions of some sort). The observations at the 1842 eclipse initiated a discussion of the physical nature of these solar appendages which continued throughout the century. The prominences (usually referred to in the literature of the time as "protuberances" or "flames") caused least dissension. They were, by general agreement, considered to be clouds floating above the solar surface. The nature of the corona was a much more controversial point. Most, but not all, observers agreed that it was not due to solar irradiation of a lunar atmosphere (by the first half of the nineteenth century few astronomers still believed that the Moon had an appreciable atmosphere). Nor, on the other hand, did the majority believe it to be due entirely to a solar atmosphere. Instead, diffraction of light round the solar limb, or scattering in the Earth's atmosphere, were invoked.

The existence of a continuous chromospheric band all round the solar limb was only noticed by a few observers at the 1842 eclipse. However, another eclipse, this time visible from northern Europe, occurred in 1851. Several observers at this eclipse commented on the appearance of the chromosphere, particularly on its jagged boundary. G. B. Airy (1853), the Astronomer Royal, was led by this to call it a "sierra", a name which continued to be used for the chromosphere by some writers until late in the century. Several prominences were noted at the 1851 eclipse, and their observation revealed the significant fact that they lay, for the most part, over regions of the Sun's surface occupied by spots. It was also observed that the Moon passed in front of the prominences as it moved across the face of the Sun. This was generally accepted as good evidence for their solar origin, although some quite able astronomers remained unconvinced. The French astronomer H. A. E. A. Faye, for example, still inclined to the belief that prominences were optical phenomena in some way connected with the Moon.

The first attempt to photograph an eclipse was made by an Austrian, Majocchi, using an early daguerreotype (see: Majocchi, 1879). He was unsuccessful, but in 1851, Berkowski, a professional photographer, succeeded in obtaining an image of the totally eclipsed Sun, using a more sensitive daguerreotype (see: de Vaucouleurs (1961), Plate 2). This showed the inner corona together with some prominences. Photography became increasingly important at later eclipses, since it could capture the fleeting phenomena in a way that visual impressions could not.

The existence of dark divisions in the solar spectrum was first noted, almost casually, by W. H. Wollaston (1802).

"If a beam of daylight be admitted into a dark room by a crevice $\frac{1}{20}$ of an inch broad, and received by the eye at a distance of 10 or 12 feet, through a prism of flint-glass, *free from veins*, held near the eye, the beam is seen to be separated into the four following colours only—red, yellowish green, blue and violet. . . .

"The line A that bounds the red side of the spectrum is somewhat confused, which seems in part owing to want of power in the eye to converge red light. The line B, between red and green, in a certain position of the prism, is perfectly distinct; so also are D and E, the two limits of violet. But C, the limit of green and blue, is not clearly marked as the rest, and there are also on each side of this limit other distinct dark lines—*f* and *g*—either of which, in an imperfect experiment, might be mistaken for the boundary of these colours."

His deduction from this was simply that visible sunlight could be separated into four fundamentally distinct colours. There was no full investigation of the lines until the following decade when Joseph Fraunhofer (1817) independently noted their existence.

"In the window-shutter of a darkened room I made a narrow opening—about 15 seconds broad and 36 minutes high—and through this I allowed sunlight to fall on a prism of flint-glass which stood upon the theodolite. . . . I wished to see if in the color-image from sunlight there was a bright band similar to that observed in the color-image of lamplight. But instead of this I saw with the telescope an almost countless number of strong and weak vertical lines, which are, however, darker than the rest of the color-image; some appeared to be almost perfectly black."

Fraunhofer then went on to discuss the lines in detail. He denoted the most striking features by letters of the alphabet— from A, which approximately marked the red end of the spectrum, to I, which marked violet end.

"In the space between B and H there can be counted, therefore, about 574 lines, of which only the strong ones appear on the drawing. The distances apart of the strongest lines were measured by the theodolite and transferred according to scale directly to the drawing; the weak lines, however, were drawn in, without exact measurement, simply as they were seen in the color-image.

I have convinced myself by many experiments and by varying the methods that these lines and bands are due to the nature of sunlight, and do not arise from diffraction, illusion, etc."

Fraunhofer, himself, was mainly interested in the lines for the use they could be put to in designing more accurate optical equipment. (This is clearly revealed by the title of his paper containing an account of their discovery: *Determination of the Refractive and the Dispersive Power of Different Kinds of Glass, with Reference to the Perfecting of Achromatic Telescopes.*)

Despite a continuing interest in the Fraunhofer lines, there was still no agreement at the mid-century on where, or how, they originated. One suggestion was that they were caused by optical defects in the telescope; another supposed that they were formed in the atmosphere of the Earth. Finally, there was the obvious possibility that the lines originated in the Sun.

In 1833 Sir David Brewster examined the dark lines with a considerably larger dispersion than Fraunhofer had used. He noted that some lines were certainly produced in the Earth's atmosphere, for they darkened as the Sun declined towards the horizon. He believed, however, that the majority originated in the Sun. He concluded from laboratory experiments that the lines could be explained by absorption in a solar atmosphere, and that the apparent surface of the Sun was probably a solid raised to incandescence by intense heat (Brewster, 1834). His fellow-countryman, J. D. Forbes, initially agreed with this concept. A subsequent experiment, however, led him to reverse his opinion completely. He argued that, if the Fraunhofer lines were formed by absorption in the solar atmosphere, then they should appear darker towards the solar limb, for there they would have to pass through a greater depth of the atmosphere. To test this hypothesis, he compared visually the spectrum from the solar limb, visible at the annular eclipse of the Sun in 1836, with the normal spectrum seen out of eclipse (Forbes, 1836). He found no obvious difference between the two, and therefore concluded that the Fraunhofer lines could not be formed on the Sun. This observation continued to puzzle nineteenth-century astronomers, although Ångström subsequently claimed, from a direct comparison of the spectrum at the centre and limb of the Sun, that the fainter Fraunhofer lines did, in fact, become more clearly defined toward the limb (see: Lockyer (1874), p. 205).

Curiously enough, the crucial experiment which indicated not only the place of origin of the Fraunhofer lines, but also how they originated, had already been made by the mid-century, although its significance remained unrecognized. To understand the experiment we must first consider one of the strongest lines mapped by Fraunhofer which lay in the yellow region of the spectrum and was labelled by him the D line. It was noted that the D line really consisted of two lines very close together, which were therefore labelled D_1 and D_2 (the latter having the shorter wavelength).

These lines in emission had already been observed in some detail in the laboratory. Fraunhofer, himself, remarked on the similarity of their position in solar and terrestrial spectra. It is, in fact, possible to suppose that the very importance of the D lines exerted a retarding influence on the growth of spectroscopy. The problem was their ubiquity: they seemed to occur whenever any substance at all was burnt. Hence, they appeared to confute the otherwise reasonable speculation that different elements produced different spectral lines (which was a necessary basic premise for any system of spectrum analysis). By the 1840's, however, it was gradually coming to be accepted that the constant presence of the D lines was due to a very slight contamination of most ordinary substances by salt: the D lines were, in fact, linked with the occurrence of sodium in the sample. (It is, perhaps, significant that the precision of the spectroscope was considerably enhanced during this same period. It was, for example, during the 1840's that collimators came into general use.)

The crucial experiment was carried out by Léon Foucault in 1849. He was investigating the spectrum formed by an electric arc between carbon poles, and the inevitable D lines made their appearance. Foucault remarks:

"As this double line recalled, by its form and situation, the line D of the solar spectrum, I wished to try if it corresponded to it, and in default of instruments for measuring the angles, I had recourse to a particular process.

I caused an image of the sun, formed by a converging lens, to fall on the arc itself, which allowed me to observe at the same time the electric and the solar spectrum superposed; I convinced myself in this way that the double bright line of the arc coincides exactly with the double dark line of the solar spectrum.

This process of investigation furnished me matter for some unexpected observations, It proved to me, in the first instance, the extreme transparency of the arc, which occasions only a faint shadow in the solar light. It showed me that this arc placed in the path of a beam of solar light, absorbs the rays D, so that the above-mentioned line D of the solar light is considerably strengthened when the two spectra are exactly superposed. When, on the contrary, they jut out one beyond the other, the line D appears darker than usual in the solar light, and stands out bright in the electric spectrum which allows one easily to judge of their perfect coincidence. Thus the arc presents us with a medium which emits the rays D on its own account, and which at the same time absorbs them when they come from another quarter.

To make the experiment in a manner still more decisive, I projected on the arc the reflected image of one of the charcoal points, which, like all solid bodies in ignition, gives no lines; and under these circumstances the line D appeared to me as in the solar spectrum" (see: Lockyer, 1874, p. 186).

Experiments very similar to these later guided Kirchhoff to a general understanding of the solar spectrum—how and where it was formed—but Foucault failed to generalize his results: although they aroused some interest at the time, they were not developed further.

In England, more than one investigator skirted the idea of spectrum analysis without actually grasping it. Thus W. H. Miller (1833) passed sunlight through luminous vapours and examined the resulting spectrum. Talbot (1834) showed that flame spectra could be used to distinguish between different substances. Wheatstone (1835) concluded that the spectrum produced by a spark depended on the metal composing the electrodes. On the theoretical side, Stokes, in 1852, suggested ideas very similar to those eventually reached by Kirchhoff. He mentioned them in conversation to Lord Kelvin, who introduced them into his lectures at Glasgow University during the 1850's. All these represent partial anticipations of Kirchhoff's fundamental work on spectrum analysis published around 1860. None of them, however, provided a complete picture as he did. It is interesting in this connection to quote Stokes' opinion, as one of the people involved (see: Roscoe (1907), p. 70).

"As well as I recollect what passed between Thomson [i.e. Lord Kelvin] and myself about the lines was something of this nature. I mentioned to him the repetition by Miller of Cambridge of Frauenhofer's observation of the coincidence of the dark line D of the solar spectrum with the bright line D of certain artificial flames, for example a spirit lamp with a salted wick. Miller had used such an extended spectrum that the 2 lines of D were seen widely apart, with 6 intermediate lines, and had made the observation with the greatest care, and had found the most perfect coincidence. Thomson remarked that such a coincidence could not be fortuitous, and asked me how I accounted for it. I used the mechanical illustration of vibrating strings which I recently published in the *Phil. Mag.* in connection with Foucault's experiment. Knowing that the bright D line was specially characteristic of Soda, and knowing too what an almost infinitesimal amount suffices to give the bright line, I always, I think connected it with soda. I told Thomson I believed there was vapour of Sodium in the sun's atmosphere. What led me to think it was *sodium* rather than soda, chloride of sodium, etc., was the knowledge that gases that absorb (so far as my experience went) yield solutions that absorb in the same general way, but without the *rapid* alterations of transparency and opacity. Now if the absorption were due to vapour of chloride of sodium we should expect that chloride of sodium and its solution would exercise a general absorption of the yellow part of the spectrum, which is not the case. Thomson asked if there were any other instances of the coincidence of bright and dark lines, and I referred to an observation of Brewster's relative to the coincidence of certain red lines in the spectrum of burning potassium and the lines of the group *a* of Frauenhofer. . . . Thomson with his usual eagerness said, oh then we must find what metals produce bright lines agreeing in position with the fixed dark lines of the spectrum, or something to that effect. I was, I believe rather disposed to rein him in as going too fast, knowing that there were terrestrial lines (seen when the sun is low) which evidently take their origin in terrestrial atmospheric absorption where metals are out of the question, and thinking it probable that a large number of lines in the solar spectrum might owe their origin to gaseous absorption of a similar character in the solar atmosphere. . . .

The idea of connecting the bright and dark lines *by the theory of exchanges* had never occurred to me, and I was greatly struck with it when I first saw it, which was in a paper of Balfour Stewart's read before the Royal Society and printed in the Proceedings. I was wrong in saying *lines*, for B. Stewart considers only solids, the spectra of which don't present such abrupt changes. Stewart's paper was independent of, but a little subsequent to, Kirchhoff's, though the same idea with reference to radiant heat occurs in two papers of his printed in the *Edin. Phil. Trans.* and much anterior to Kirchhoff's paper. These papers I was not acquainted with at the time when Stewart's paper on light came before the R.S."

Perhaps A. J. Ångström of Uppsala came nearest to anticipating Kirchhoff. In an article published in 1853 he discussed several of the fundamental ideas of spectrum analysis. He tried, for example, to show that a hot body emits the same kind of radiation it absorbs. He also emphasized the physical relationship which must exist between some of the dark Fraunhofer lines in the solar spectrum, and the bright lines found in the laboratory spectra of certain metals. Unfortunately, the work was published initially in the *Transactions of the Royal Academy of Stockholm* in his native language, and did not receive the attention it deserved.

In 1842, Christian Doppler, a Professor at Prague, pointed out that the motion of a body could affect the colour of the light emitted. Doppler's theory, in the form he put it forward, was partly erroneous, and, in any case, was not immediately applicable to astronomical observations. However, in a talk given in Paris in 1848, Fizeau pointed out that the presence of Fraunhofer lines provided a delicate test for the existence of radial motion, if Doppler's ideas were correct. This talk was not published in full at the time, but the possibility of detecting the Doppler effect in this way was being discussed amongst astronomers during the 1850's.

The existence of infrared radiation had first been established by Sir William Herschel (1800) right at the end of the eighteenth century. The discovery resulted from an investigation into the heating power of different parts of the solar spectrum: he observed that maximum heating occurred beyond the end of the red. Not long afterwards, J. W. Ritter (1801), inspired by Herschel's work, investigated the action of different parts of the solar spectrum on silver chloride. He found that maximum reduction occurred beyond the violet end of the visual spectrum, thus revealing the presence of ultraviolet radiation. Herschel carried out a considerable number of experiments on infrared radiation which led him to conclude that radiant heat was fundamentally different in its nature from visible light. Subsequently, Sir David Brewster extended this proposition to the conclusion that infrared radiation, visible light and ultraviolet radiation were all basically different, although they were found mingled together in sunlight. This idea continued to appear until quite late in the nineteenth century. It was dispelled partly by the growing observational evidence that the different types of radiation merged into each other. The major advances in correlating the ultraviolet with the visible up to the mid-century were made by Sir John Herschel (1840), A. E. Becquerel (1842) and J. W. Draper (1842), all of whom managed to photograph the solar spectrum in the early 1840's. Like the majority of nineteenth-century photographs these were more sensitive at the blue end of the spectrum than the human eye. As a result, they showed absorption, previously unobserved, in the ultraviolet, which appeared to be similar to that already found in the visible. A few years earlier, Sir John Herschel (1840) had similarly exposed a strip of black paper moistened with alcohol to radiation from beyond the red end of the visual spectrum. He found that the alcohol dried unequally, thus suggesting the existence of something like Fraunhofer lines in the infrared radiation. Both of these experiments indicated the essential similarity of the visible and the invisible radiations.

The belief that energy must be conserved began to establish itself during the 1840's. By this time the geological evidence for an age of the solar system much longer than the traditional 6000 years was well established. As a result, the source of solar energy came under discussion. It quickly became evident that chemical burning was quite inadequate to account for the solar radiation over any prolonged period of time. One of the people who made this calculation was J. R. Mayer (1848), an early pioneer in the establishment of the law of energy conservation. It was becoming known at about this time that interplanetary space contained a large number of meteors (or meteorites—these words were hardly distinguished in the nineteenth century). Mayer therefore suggested that the Sun's heat was caused by the impact of these interplanetary particles on the solar surface. It was recognized from the start that this explanation encountered a major difficulty: the amount of impacting material required to account for the Sun's heat was so large that the mass of the Sun would be increasing at an appreciable rate. This, in turn, would alter the Earth's orbit and, therefore, the length of the year to an improbable extent. Mayer claimed that the gain in mass was balanced by an equivalent loss due to the mass carried away as Newtonian light corpuscles, but this idea received little support.

In 1854 Hermann von Helmholtz suggested in a popular lecture (not published until some years later) that the Sun could gain heat from its gravitational energy by contracting. (The relevant part of this talk is reprinted in Part 2.) This mechanism dominated subsequent nineteenth-century thought on the subject. It could be shown easily that the rate of contraction required was well below the observable limit. The lifetime so derived for the Sun was several million years which, at the mid-century, appeared fairly reasonable—at least to astronomers.

CHAPTER II

The New Astronomy (1850-1900)

Solar Spectroscopy

The final steps in the development of spectrum analysis and its application to the Sun were taken by Gustav Kirchhoff in Heidelberg (working initially in conjunction with Robert Bunsen). Their first communication on the subject was sent to the Berlin Academy at the end of 1859, but the work developed continuously over the next two years (see the papers reprinted in Part 2). Two main conclusions were essentially reached. In the first place, it was found that incandescent solids, or liquids, gave continuous spectra, whilst gaseous spectra consisted either of bright lines, or of bands; (the wavelengths being characteristic of the gas). Secondly, it was found that, when light from an incandescent solid, or liquid, passed through a gas, the wavelengths at which it was absorbed, were the same as those at which the gas characteristically emitted, if heated to the same temperature. These two propositions, although subsequently somewhat modified, between them summed up the basis of spectrum analysis. We have seen that in the forties and fifties there were several near anticipations of these results. We have mentioned, for example, that the second conclusion described above was also suggested by Balfour Stewart for infrared radiation in 1858, and he extended it to visible radiation two years later whilst still unaware of Kirchhoff's work. Where Kirchhoff and Bunsen scored, however, was in their very careful experimental techniques which made their conclusions immediately acceptable. Balfour Stewart was, moreover, mainly interested in continuous radiation. He did, however, point out the significance of the sodium D lines.

Kirchhoff used his laboratory results to derive a theory of the solar constitution (the first to have a thoroughly physical basis). He suggested that the photosphere of the Sun was a hot, incandescent liquid, which therefore emitted a continuous spectrum. (The observed motions of the solar surface certainly precluded the possibility that it was a hot solid.) Above the photosphere there was an extensive gaseous atmosphere (which Kirchhoff identified with the corona) and this produced the Fraunhofer lines. He ignored the Wilson effect, returning instead to the older idea that sunspots were clouds which he assumed to be floating in the solar atmosphere just above the liquid surface. This general picture received relatively little support from astronomers: although Gustav Spörer subsequently pressed for its acceptance. On the other hand, Kirchhoff's mode of approach to the problem was accepted, and this cast very considerable doubts on the Herschelian belief in a dark central sphere. Instead, it became more widely accepted that the Sun must be hot throughout (as had, indeed, already been supposed by some astronomers). The obviously high temperature of the Sun made it very difficult to accept Kirchhoff's picture of a liquid sphere. Shortly after Kirchhoff's work appeared, both Angelo Secchi (1864) and Sir John Herschel (1864) concluded that the photosphere must represent a layer of cloud in an otherwise mainly gaseous Sun. This belief was held with little dissent for the next half-century, although it was soon realized that the exact nature of the cloud particles raised problems: at the sort of temperature estimated for the Sun, any known terrestrial substance should be vaporized. G. J. Stoney (1867) suggested that incandescent carbon grains were involved, and this was widely accepted, but particles of silicon or boron and droplets of molten metal also received some support.

The first really detailed map of the solar spectrum was also produced by Kirchhoff (1861, 1862) as a concomitant part of his work on spectrum analysis. He employed a four-prism spectroscope, giving a reasonably high dispersion, and measured the wavelengths of the lines with a micrometer. At the same time, he made eye-estimates of the line intensities. (It must be remembered that all the early spectroscopic work, whether astronomical or laboratory, relied on visual measurements.) From these data he drew up a map of the visual solar spectrum. The strain of working at very low light levels so affected Kirchhoff's eyesight that the map had to be finished off by K. Hofmann, one of his pupils. Kirchhoff examined the central region of the visible spectrum between the Fraunhofer A and G lines; Hofmann extended these observations both towards the red and towards the blue. The resultant map suffered, however, from an essential defect: the dispersion was prismatic, and the scale therefore varied with wavelength. Kirchhoff actually added a quite arbitrary scale to his drawings, running from small numbers at the red end of the spectrum to large numbers at the blue. (The Fraunhofer A line was at 405 on this scale, the D lines were at just over 1000.) The difficulty of reproducing this system prevented it from ever gaining complete acceptance. Some of the more important lines were, nevertheless, often referred to by their Kirchhoff number long after our present-day wavelength system had been introduced, and this can some-times cause confusion.

A few years later, A. J. Ångström (working partly in conjunc-tion with T. R. Thalén) produced a new map of the solar spectrum (1869). Unlike Kirchhoff, he employed a grating, with the result that his map had normal dispersion. He was therefore able to calibrate his wavelengths in terms of the standard metre, instead of using an arbitrary scale. The unit he introduced of one ten-millionth of a millimetre (later called after him) received general acceptance, but his map came to be regarded after a few years as somewhat defective (mainly because of imperfections in his original grating). Consequently H. A. Rowland started on a new map of the solar spectrum (this time photographic) in the 1890's. Rowland, using one of the very fine gratings for which he was famous, mapped the solar lines in the range 3000–7000Å. (The dispersion he employed was so high that his final map had a total length of over 40 feet.) He measured the wavelengths of some 20,000 lines and gave eye-estimates of their intensities, claiming that

". . . the present series of standards can be relied on for relative wave-lengths to ·02 division of Ångström in most cases, though it is possible some of them may be out more than this amount, especially in the extreme red" (Rowland, 1889).

His tables (subject to some revision) remain in use today.

The effort put into mapping the solar spectrum during the nineteenth century was motivated to a great extent by the wish to identify the elements present on the Sun. This work was again initiated by Kirchhoff. He began by comparing the simple spectrum of sodium, as seen in the laboratory, with the solar lines. Subsequently he progressed to the extremely complicated spectrum of iron, finding nearly fifty coincidences in wavelength between the solar and the laboratory lines. As he pointed out, the odds were very greatly against this being a chance agreement. Kirchhoff used a spark source to obtain his laboratory spectra. Ångström in his work employed an arc source which gave a brighter light than the spark, and therefore allowed him to use a higher dispersion. The spark and the arc remained the two standard methods for obtaining comparison spectra throughout the nineteenth century. Kirchhoff found line coincidences for the following elements: sodium, calcium, barium, strontium, magnesium, copper, iron, chromium, nickel, cobalt, zinc and gold. His assistant Hofmann, subsequently extended the list. It was recognized, however, that line coincidence by itself was not the only criterion for affirming the presence of an element on the Sun, other factors, e.g. line intensities, had to be taken into account. Moreover, some of these early identifications depended on the coincidence of only one or two lines, and were therefore very dubious. Nevertheless, when Ångström and Thalén subsequently extended these investigations they found it possible to increase the number of line coincidences for several of the elements identified by Kirchhoff, and to find further coincidences for other elements. (The most important new element found was hydrogen.) A factor which had considerable influence on contemporary thought was that all these early lists of line identifications contained only metals (in view of its chemical activity, hydrogen was usually classified along with the metals). Even towards the end of the century, when Rowland could publish a list of some 36 elements probably present on the Sun, the huge majority were still metals. The most interesting exception was carbon. Bands attributable to some form of carbon compound were—so Lockyer believed—observed by him in the solar spectrum during the 1870's (Lockyer, 1878). Although this identification was initially disputed, evidence accumulated until by the end of the 1880's the presence of carbon on the Sun was generally accepted. The discovery was rather embarrassing for some contemporary lines of thought since, according to them, the Sun's surface should have been too hot for the survival of molecules.

So far we have been discussing the identification of elements from their absorption lines. It was also claimed during the nineteenth century that some elements were represented in the solar spectrum only by emission lines. The most famous of these was helium. As will be described later, the existence of helium in the solar chromosphere first became apparent at the eclipse of 1868. During this eclipse several observers noted a chromospheric line in the yellow. Most of them identified it as the sodium D line in emission, but Janssen pointed out that there was a discrepancy in the wavelength. This difference was also soon picked up by Lockyer observing the chromosphere out of eclipse. (The papers by Janssen and Lockyer are reprinted in Part 2.) Lockyer decided that the new line (which came to be called D_3) could not be ascribed to any known terrestial element. It was therefore suggested that a new element—later called helium—had been detected. Identification of a new element on the basis of one line only was greeted with some reserve, but other lines which could also be attributed to helium were subsequently discovered, and the element gradually acquired respectability.

Down the years there were various claims to have identified helium in terrestrial materials, but the first genuine identification was delayed until 1895 when Ramsay, in examining the gases given off by a mineral called clevite, detected the presence of the D_3 line.

"Prof. Miers, in a letter which he wrote me the day after an account of the fruitless attempts to cause argon to combine had been given to the Royal Society, drew my attention to experiments by Dr. Hillebrand of the United States Geological Survey, in course of which he obtained a gas, which he believed to be nitrogen, by treating the rare mineral clevite, a substance found in felspathic rocks in the south of Norway, with sulphuric acid. The chief constituents of clevite are oxides of the rare elements uranium and thorium, and of lead. The gas obtained thus, after purification from nitrogen, was examined in a Plücker tube with the spectroscope and exhibited a number of brilliant lines, of which the most remarkable was one in the yellow part of the spectrum, similar in colour to the light given out by the glowing tube. The position of this line, and of others which accompany it, established the identity of this gas, not with argon, as was hoped, but with a supposed constituent of the Sun's chromosphere" (Ramsay, 1901).

A subsequent study enabled Ramsay to separate helium in a fairly pure state. The detection of helium on Earth, however, did not immediately bring an end to all confusion. From a detailed laboratory examination of the helium spectrum, Runge and Paschen (1895) concluded that it could be divided into two quite distinct sets of lines. They therefore suggested that helium was really a mixture of two different gases, and, from the relative heights of appearance of the lines in the solar atmosphere, they deduced that one of the gases was slightly denser than the other. Several attempts were made in the next few years to separate these two gases, but, needless to say, they failed completely and the whole concept was fairly soon dropped. It was also pointed out that the D_3 line observed in the laboratory was not single, as astrophysicists had always supposed, for it had a faint companion. For a short time this cast doubt on the similarity of the solar and terrestrial lines, but soon several observers managed to find the fainter component in the D_3 emission from prominences, where it had previously been overlooked.

At the solar eclipse of 1869, C. A. Young (paper reprinted in Part 2) and, independently, W. Harkness (1867 [sic]) both discovered a new emission line, this time in the solar corona. It was eventually agreed that this line (usually referred to in nineteenth-century literature as the 1474 line from its wavelength on the Kirchhoff scale) was not due to any known terrestrial element. By analogy with helium, it was therefore ascribed to another hypothetical element—coronium. Belief in the existence of this element lingered on for many years—well into the present century —until it was finally obvious that there was no place for it in the periodic table, and it must therefore be, not an unknown element, but a known element in an unusual state. During the many years of its acceptance there were, of course, several attempts at identification. It was, in particular, widely believed that coronium and helium must be related. Thus, Liveing and Dewar (1879) suggested that a connection existed between the 1474 line and two of the helium lines. Later, when Runge suggested that helium contained two components, it was thought that one might be helium and the other coronium. This hope was rapidly dashed by the non-appearance of the coronium line in the spectrum of terrestrial helium. Nevertheless, it was commonly supposed in the early years of the present century that coronium would prove to be one of the rare gases.

Towards the end of the 1870's, the American astronomer, Henry Draper, announced that he had discovered bright lines due to oxygen in the Sun (Draper, 1877). It was suspected from the beginning that these so-called emission lines were really just places where the bright continuous radiation of the Sun shone through owing to an absence of absorption lines. However, Draper repeated his observations at a considerably higher dispersion and still found a good coincidence with the spark spectrum of oxygen. It was therefore widely held for some years that oxygen, like helium and coronium, was to be seen on the Sun predominantly in emission. Subsequently observations at still higher dispersion were made which revealed that the coincidences found by Draper were not exact: wavelength differences existed between the bright "lines" in the Sun and the laboratory spectra. Thus, by the end of the nineteenth century, belief in the existence of bright oxygen lines had virtually disappeared. About this time, however, oxygen was positively identified in the Sun—by Runge and Paschen (1897)—as an absorption triplet in the red. In fact, oxygen absorption had been found in the solar spectrum earlier than this, but it had always previously been due to terrestrial rather than solar oxygen.

The presence of terrestrial absorption lines in the solar spectrum seriously complicated the early spectroscopic investigations of the Sun. As we have seen, Brewster in his early investigations of the solar spectrum noted the presence of lines formed in the Earth's atmosphere. This work was continued by Piazzi Smyth (1858), then Astronomer Royal for Scotland, who made a series of observations of the solar spectrum from Teneriffe in the Canary Islands. Around 1860, Brewster returned to the investigation. It is an odd fact that by the time that Kirchhoff's work was confirming Brewster's earlier belief in the solar origin of many of the Fraunhofer lines, Brewster, himself, doubted it. Working now with J. H. Gladstone, he first suggested that the solar lines were due to some form of interference (1860). Next they considered how many of the lines might reasonably be attributed to elements in the atmosphere of the Earth. They identified a considerable number of lines and bands in the solar spectrum whose intensity depended on the altitude of the Sun, but they were unable to determine the atom or molecule involved. In his attempts to identify the lines, Gladstone examined a terrestrial source of light (Beachy Head lighthouse) from several miles away, but was unable to detect any atmospheric lines. A few years later the French astronomer, P. J. C. Janssen, tried a similar experiment first at Geneva, and afterwards at Paris and was more successful (Janssen, 1865). As a result he was able to identify water vapour as the major factor in producing terrestrial absorption lines. Later Ångström (1867) pointed out that certain of the terrestrial lines—in particular those denoted as A and B by Fraunhofer—did not vary with climatic conditions as the majority of the terrestrial lines did. He therefore supposed that they were caused by some other molecule—possibly carbon dioxide. Eventually Egoroff (1883) managed to show that these lines were actually due to terrestrial oxygen. About this time an easy way of separating terrestrial from solar lines appeared. S. P. Langley (1877) devised a method for comparing the spectra from the opposite limbs of the Sun, and found on examination that the two spectra were obviously displaced relative to each other. This was a result of Doppler shift due to solar rotation and, therefore, did not affect the terrestrial lines. The test was subsequently applied by M. A. Cornu (1884) to distinguish between solar and terrestrial lines, and soon became the standard method.

The presence of a Doppler shift at the limb due to solar rotation was first detected by H. C. Vogel in 1871 (paper reprinted in Part 2), using an instrument of a type devised by J. K. F. Zöllner. Preliminary estimates of the magnitude of the displacement were made by Young (1876) using the first successful astronomical spectroscope with a grating as the dispersive element. The first accurate and extensive measurements were, however, made by N. C. Dunér at the end of the eighties. Dunér (1891) used a relative method of measurement based on the difference in behaviour of the solar and terrestrial lines. He took a pair of iron lines in the red part of the spectrum and compared their wavelengths with two nearby terrestrial oxygen lines. His results confirmed those already obtained from direct observations on the motion of sunspots (to be discussed later), but the coverage of latitude was double that of the direct measurements. Another very extensive series of solar rotation measurements were carried out by J. Halm at Edinburgh in the early years of the present century (for a general survey see Halm, 1922). He concluded from his observations that the solar rotation rate might be variable from year to year. Halm's survey, like all its predecessors, was based on visual measurements. The difficulty of detecting small line-shifts in this way, however, led inevitably to all subsequent work in this field being photographic.

Observation of the Solar Surface

In the early 1860's, a major controversy developed over the nature of the surface markings on the Sun. James Nasmyth (1862) claimed that he could resolve the markings on the solar surface into separate bright objects, each having an appearance rather similar to that of a willow leaf. He was firmly contradicted in the following year by Rev. W. R. Dawes (1863):

"The [solar] surface is principally made up of luminous masses imperfectly separated from each other by rows of minute dark dots—the intervals between these being extremely small and occupied by a substance decidedly less luminous than the general surface. This gives the impression of a division between the luminous masses, especially with a comparatively low power, which, however, when best seen with high power, is found to be never complete. The masses thus incompletely separated are of almost every variety of irregular form, the rarest of all, perhaps, being that which is conveyed to my mind by Mr. Nasmyth's appellation of '*willow-leaves*', i.e. *long, narrow, and pointed*. Indeed, the only situation in which I have usually noticed them to assume anything like such a shape is in the immediate vicinity of considerable spots, on their *penumbrae* frequently projecting beyond it irregularly for a small distance on to the umbra."

Sir John Herschel had earlier compared the appearance of the solar surface with that of a flocculent chemical precipitate, seen from above as it slowly sank through a transparent fluid. Dawes was in general agreement with this description, but Herschel, himself, was prepared to countenance the willow-leaf concept. However, whereas Nasmyth had regarded the leaves as clouds, similar to those on Earth, Herschel suggested that they might be filamentary structures of solid material. (He even speculated that they might be regarded as organisms of some peculiar and amazing kind.)

"They strongly suggested the idea of solid bodies sustained *in equilibrio* at a definite level (determined by their density) in a transparent atmosphere *passing by every gradation of density from that of a liquid to that of the rarest gas* by reason of its heat and the enormous superincumbent pressure (as in the experiments of M. Cagniard de la Tour on the vaporization of liquids under high pressure); their luminosity being a *consequence* of their solidity;

transparent and colourless fluids radiating no light from their interior however hot" (J. F. W. Herschel, 1869, p. 696).

Notice here the emphasis on high pressure: it was often assumed at the time that, since the Sun was much more massive than the Earth, its atmospheric pressure might be considerably greater. In 1869, the year in which this edition of Herschel's book was published, Andrews showed that there existed a critical temperature for any vapour above which it could not be liquefied by pressure alone. This was accepted as confirming the idea, evolved in the 1860's, of a mainly gaseous Sun whose gas content nevertheless sometimes attained the density and consistency of a liquid.

Several years earlier, Arago had investigated the possible occurrence of polarization in sunlight, but had found no sign of it anywhere on the solar disc. This led him to argue that the atmosphere of the Sun must be wholly gaseous, since if solid or liquid particles were present they would produce partial polarization near the limb. This conclusion was accepted by some later workers, but was criticized by Sir John Herschel who suggested that it was only applicable to a smooth surface (for a discussion see: J. F. W. Herschel, 1869, pp. 245 f.). Since the Sun's surface was manifestly rough, he deduced that no overall polarization need appear.

The controversy over the appearance of the solar surface continued to bubble throughout the sixties. Secchi, and others, described the surface markings as being more like rice grains; Huggins sketched apparently intermediate forms. It was agreed in the end that a typical granule (the word was introduced by Dawes) was oval in shape, about 1″ wide and 1·5″ long, but near sunspots it could become much more elongated. The dispute had the beneficial effect of standardizing the terminology used in describing the solar surface. It also had the result that attention was focused on the Sun: Lockyer's early solar observations were a direct consequence of this altercation.

During the 1870's, the solar granulation was investigated in particular detail by Langley in the United States and Janssen in

France. Langley (1874), under very favourable seeing conditions, observed some granules which were as little as 100 miles across the figure of 1″, mentioned above, corresponding to about 500 miles). He estimated that the bright granules gave out some three-quarters of the total light from the Sun, but they took up only one-fifth of the surface area. (Janssen considered their emissivity to be even greater.) Langley believed that bright granules represented the points where hot currents were ascending from the interior of the Sun, whereas the dark interstices represented regions where cooler currents were descending. This picture received widespread assent during the latter part of the nineteenth century.

All the early work on solar surface markings was visual, the intensity of the sunlight being reduced by filters or poorly reflecting surfaces inserted in the telescope. In 1854, J. B. Reade in England used a dry collodion plate to obtain a solar image showing the mottled structure of the Sun's surface. The real use of photography for the investigation of granulation was, however, initiated by the French astronomer, Janssen, in the 1870's (paper reprinted in Part 2). He used a telescope specially constructed for solar photography; it provided very short exposure times and a large image. The photographs he obtained with this instrument remained the best delineations of granular structure for well over half a century. At the time, however, the main interest of his photographs was less in the granules themselves, than in the hazy and ill-defined nature of the granulation which appeared over certain regions of the solar surface. It was realized that this might be due simply to air currents in the Earth's atmosphere, or in the tube of the telescope (this is the explanation accepted nowadays). Janssen, however, believed that photographs taken in quick succession showed the same areas of the solar surface to be hazy. He therefore asserted that it was an effect in the solar atmosphere, and named it the *réseau photosphérique*. It was supposed to appear whenever the solar atmosphere was particularly disturbed.

Sunspots

Shortly after the middle of the nineteenth century, R. C. Carrington, a wealthy English amateur, began a detailed study of sunspots from his private observatory at Redhill. He had become interested in the relationship, then recently announced, between the sunspot cycle and the geomagnetic fluctuations. His examination of the sunspot records had convinced him that they were inadequate for tracing such relationships in detail, so he had decided to make his own observations covering one complete sunspot cycle. As it happened, he was forced to cut his observing programme short (owing to increasing business commitments). Nevertheless, he obtained excellent results for the period 1853–61. At this time Sir John Herschel was urging very strongly that there should be daily photographic records of the Sun's surface. Carrington, however, felt, when he started, that solar photography was still at too experimental a stage to make it worth while for routine observations. He therefore relied entirely on the visual recording of a projected solar image.

One of the main items in Carrington's investigation was the rotation rate of the Sun, as derived from the apparent motions of spots (Carrington, 1863). The rotational period of the Sun was still poorly determined at the mid-century despite several attempts. Carrington's long series of observations at last indicated the cause of the difficulties: the Sun's surface rotated differentially (that is, the period of rotation depended on the solar latitude). His analysis suggested that the period depended on $\sin^{7/4}\phi$ (where ϕ is the solar latitude). This was a purely empirical formula with no theoretical justification. Subsequent investigators therefore tended to prefer a sine squared law which could be justified theoretically. Indeed, the French astronomer, Faye, re-examined Carrington's results and claimed that they could be represented almost equally well in this way. Carrington also decided, when finally summarizing his results, that the latitude with the maximum rate of rotation was not the solar equator, but one degree further north. Here, too, he found few followers.

As important as the discovery of the equatorial acceleration of the Sun, was Carrington's clarification of the variation of the sunspot distribution over the solar surface during the course of a spot cycle (see the paper reprinted in Part 2). He described in detail how the spots from the old cycle finally disappeared at low latitudes, whilst the new spots were appearing at higher latitudes, and how subsequently, the spot positions worked their way downwards again. His results were soon confirmed by others, particularly the German astronomer, Spörer, who made a survey of sunspot characteristics, similar to Carrington's, throughout the sixties. Spörer (1861) actually published an independent account of the equatorial acceleration after only a year's observation. Some time later, he combined all the sunspot records available to derive a statistically more certain result for the variation of rotation with solar latitude. He supposed, however, a simple sinusoidal variation with latitude.

Knowledge of the equatorial acceleration from sunspot motions was necessarily limited to those latitudes where sunspots appeared (usually not much more than $\pm$ 35°). The spectroscopic observations, as has already been pointed out, were important because they enabled measurements to be made at much higher latitudes. (The spectroscopic results were usually reduced to a sine squared formula similar to that for the sunspots.) They were also important, however, because they showed that the photosphere rotated in the same way as the sunspots: a fact which had sometimes been questioned.

Towards the end of the 1850's, the Royal Society finally heeded Sir John Herschel's plea for a continuous photographic patrol of the Sun's surface. In 1857, Warren De la Rue was asked to devise a suitable instrument for this purpose to be installed at Kew. A photoheliograph was duly set up there in the following year, and after a fairly prolonged period of adjustment went into regular service. It finally ceased operation in 1872, but almost immediately a similar survey started up at Greenwich. The photoheliographic record in this country has been maintained continuously since then. The photographs produced by the patrol were, of course, mainly valuable for their record of sunspots, and De la Rue, himself, was especially concerned with research in this field. At the beginning of the 1860's, he started to produce successful stereoscopic pictures of the Sun, having found that two photographs of the Sun taken at somewhat less than half an hour apart, when combined, gave an excellent impression of depth. He was thus able to provide a direct demonstration of the relative heights above the photosphere of the different solar phenomena. He confirmed, in particular, that sunspots appeared to be regions depressed below the general level of the solar surface (De la Rue *et al.*, 1865a).

The Wilson effect was reinvestigated several times during the latter half of the nineteenth century. Besides his stereoscopic investigation, De la Rue also joined with Stewart and Loewy and examined the shape of spots as they crossed the solar surface in the traditional way. They finally concluded that the Wilson effect could be observed in about three-quarters of all sunspots (De la Rue *et al.*, 1865a). Secchi (1866) also made a detailed investigation of the effect, but in a different way. He measured the actual position of particular points in the umbra of a spot as it crossed the Sun's surface. This enabled him not only to confirm that the Wilson effect was present, but also to quote actual figures for the depth of the depression (he found that sunspots were relatively shallow). During the 1890's, however, the Wilson effect came in for some strong criticism. Rev. F. Howlett (1894) declared to the Royal Astronomical Society that the sunspot drawings he had made over many years provided little evidence for the effect. He received support from Rev. W. Sidgreaves (1895) who made a similar study of the long run of sunspot drawings at Stonyhurst. On the other hand, at almost the same time, the Italian astronomer, A. Riccò (1897), produced more evidence in favour of the effect.

It was shown by Henry and Secchi, and later by Langley's more extensive measurements, that sunspots radiated less heat than the surrounding photosphere. Other observations by Langley (1876) in the 1870's seemed to show that the radiation from the spots and the photosphere decreased in a constant ratio from the centre to the limb of the Sun. Measurements made in the 1890's by E. B.

Frost (1896) in the United States and by W. E. Wilson (1895) in Britain seemed to show, on the contrary, that the radiation from spots decreased less towards the limb than the photospheric radiation. This observation led Frost to conclude that sunspots lay above the photosphere, and so suffered less absorption towards the limb. It was suggested that the discrepancy in these observations resulted from the different epochs at which they had been made: Langley observed near a minimum of the solar cycle, Frost and Wilson near a maximum. For example, E. R. von Oppolzer (1893) had supposed that there were different amounts of chromospheric activity present at the two times. There were, however, very considerable variations in the measurements even of a single observer: Frost and Langley both observed spots which appeared to be hotter than the surrounding photosphere.

Frost (1896) noted that slight differences in the rotation rates derived from sunspots and from the photospheric spectrum also suggested that the spots lay above the photosphere. In an attempt to reconcile the conflicting evidence, some support was given to the suggestion that the discrepancies were due to variable amounts of gas within the hollow formed by the spot: this led in turn to varying refractive effects which would alter the apparent level observed. Nevertheless, it seems that the existence of the Wilson effect was regarded with greater reserve in 1900 than in 1850.

Carrington's extensive data on spot motions besides providing information on the solar rotation, also gave the first detailed indication of how spots moved relative to the photosphere. This knowledge was further extended by Spörer's results. On the basis of Sir William Herschel's model of the Sun, it had been customary to assume that the clouds forming the apparent solar surface were subject to the same sort of motions as terrestrial clouds. The new data, however, (as was emphasized by the French astronomer, Faye) showed that this simple idea was no longer tenable.

Considerable effort was expended during the nineteenth century on the study of the development and evolution of spots. Although the complexity of the subject led to a good deal of disagreement in detail, the different studies showed overall agreement on the main features. There were two points, however, which provoked continuing discussion throughout this period. The first concerned the existence of rotatory motions in and around sunspots. The second centred on the appearance of light bridges and veils across spots.

As we have seen, the Herschelian description of the solar atmosphere suggested a direct analogy between sunspots and cyclonic motions in the Earth's atmosphere. Thus Dawes, describing a spot he observed in 1852, said:

"It appeared as if some prodigious ascending force of a whirlwind character, in bursting through the cloudy stratum and the two higher and luminous strata, had given to the whole a movement resembling its own" (Dawes, 1852).

Long series of observations in the latter half of the century showed, however, that only a few per cent of all sunspots gave any clear indication of a spiral nature. The emphasis on the cyclonic basis of sunspots therefore waned. As we shall see in the next chapter, it was revived on the basis of spectroscopic measurements early in the present century. It then remained fairly popular until the 1950's.

Nineteenth-century astronomical literature contains many descriptions of bright bridges forming across sunspots (usually described in terms of the photosphere overflowing into the spot). Normally this was a relatively slow process (lasting perhaps, for a few days) which was associated with the decay of a spot. The major interest, however, was in the unusual phenomena which frequently seemed to occur during the establishment of a bridge. Some of these descriptions may have been completely accurate; others were very probably due to the use of small instruments under bad seeing conditions. As an example, which might fall into either category, we can quote C. H. F. Peters (1855) early observations (made in the 1840's) of very rapid motions in the establishment of a light bridge.

"Two of the notches from opposite sides [of the sunspot] step forward into the area, over-roofing even a part of the nucleus; and suddenly from their prominent points flashes go out, meeting each other on their way, hanging together for a moment, then breaking off and receding to their points of starting. Soon this electric play begins anew and continues for a few minutes, ending finally with the connection of the two notches, thus establishing a bridge, and dividing the spot into two parts."

As we have seen, Carrington's and Spörer's researches set fairly precise limits on the extent of the sunspot zones. Subsequently, however, E. L. Trouvelot (1876) suggested that protospots could be discerned at much higher latitudes. He named them "veiled" spots, assuming that they were basically similar to ordinary spots, but that insufficient energy was available at high solar latitudes for them to break completely through the photosphere. Not many people were able to observe the phenomena described by Trouvelot, and relatively little further work was done on them.

Observations of sunspots were often accompanied by observations of faculae. In fact, it was established quite early on that faculae, although found over much of the solar surface, were particularly abundant around sunspots. (There was a certain amount of controversy over the exact relationship between faculae and sunspots: for example, Secchi claimed that faculae appeared before the corresponding sunspots, whereas Lockyer believed that they appeared afterwards.) It was thought throughout most of the nineteenth century that they were simply elevated regions of the photosphere: rather like mountains on Earth except for their lack of solidity. This picture was confirmed by the appearance of faculae at the limb, where they stood out as projections, and De la Rue's stereoscopic photographs also showed that the faculae were above the photosphere (De la Rue, 1862). In one case, however, De la Rue observed facular material lying at a considerable height above a sunspot. This was subsequently taken as an indication that faculae might be more like clouds than mountains, a view which received some consideration late in the century.

Direct photography of faculae proved to be no easier than their visual observation. At the end of the 1880's, however, J. Wilsing

(1888) carried out a series of measurements on the motion of faculae based on photographs. His results indicated that faculae showed a similar rotation rate over a wide variety of latitudes. This was entirely unexpected, and other investigations, particularly in England and Russia, soon contradicted it. Nevertheless, the ultimate conclusion reached was still surprising: it was found that the faculae were moving more quickly than the sunspots, even though the spots certainly underlay the faculae. On the other hand, the solar rotation rate derived from the Fraunhofer lines was somewhat slower than that derived from spots. But the lines were supposed to be formed in the solar atmosphere above the level of the spots (and probably above the faculae too). These apparent discrepancies remained unresolved (although it was accepted during the early years of the present century that the solar rotation rate, in general, increased with height above the photosphere).

1866 was a significant year for solar physics. During it, Lockyer began to study the spectra of sunspots. Apart from the intrinsic importance of such work, this represented the first detailed spectroscopic examination of light from a particular region of the Sun's surface—earlier work had used light from the Sun as a whole. Lockyer turned to this field partly in the hope that it would provide a test of current ideas on the nature of spots. (These will be examined later.) He found that the spectrum of the umbra seemed to show a continuous absorption crossed by Fraunhofer lines: the lines were like those in the photospheric spectrum, but some were broader and darker in spots (paper reprinted in Part 2). At first this was disputed: Huggins (1867) claimed initially that no certain differences in the appearances of the lines could be detected. By 1868, however, he had reversed this opinion (Huggins, 1868b). At about this period, Secchi (1869b), too, was making a close examination of sunspot spectra and noting the differences from the photospheric spectrum. He decided that the spot spectrum resembled that of the solar limb. This belief was, however, short-lived, for by the early seventies he had abandoned it.

The general absorption in the spot relative to the photosphere was at first attributed, e.g. by Hastings (1881) in the United States, to an accumulation of carbon particles (or something similar) in the spot. Since the spot was thought of as a cavity filled with vapours, the resulting increase in pressure was sufficient to explain the broadened spectral lines there. In 1883, however, Young showed that the apparently continuous absorption of the spot, when seen under high dispersion, actually consisted of very many fine absorption lines close together (paper reprinted in Part 2). As he remarked, this made it certain that the absorption must be due to gases. Dunér (1891), who repeated and confirmed Young's observations some years later, pointed out that the occasional larger spaces between these lines possessed the same brightness as the photosphere. This again indicated that a source of continuous absorption, such as carbon particles, could not occur in spots. Young also noted that these fine lines generally had a spindle-shape, as did many of the Fraunhofer lines accentuated in spots. The odd way in which some sunspot lines were broadened in comparison with photospheric lines, whereas others were not, attracted considerable attention. Lockyer, in particular, made a most detailed study. In his book *The Chemistry of the Sun* published in 1887, he discussed the results from several years of observation. They led him to the conclusion that the lines which were most widened in the spectra of sunspots were not the same throughout a solar cycle. Around solar minimum the most widened lines were due mainly to various known metals. Towards maximum, however, the widest lines were mainly of unknown origin. The iron lines, in particular, changed very obviously over this period. This last observation was confirmed by Rev. A. L. Cortie (1890) at Stonyhurst. He noted, however, that the difference in appearance of the lines seemed to depend as much on the type of spot involved as on the stage in the solar cycle. Lockyer used the changes as evidence for his hypothesis that elements dissociated on the Sun (this will be discussed later in the chapter), but this had to be dropped towards the end of the century as it was then shown that many of the unknown lines were really due to vanadium and titanium.

Huggins initially thought not only that photospheric lines were not broadened in spots, but also that no new lines appeared there. This was soon shown, especially by Lockyer, to be untrue. During the 1880's, several bands were found in the spectra of sunspots, but it proved impossible to identify the molecules concerned. Some of the compounds suggested, such as iron oxide, were known to dissociate at quite low temperatures: knowledge of sunspots had grown sufficiently by the end of the century for these to be eliminated. At the end of the 1860's, Secchi thought that he could identify similar bands in sunlight as a whole when there was a layer of thin cloud in front of the solar disc. He therefore concluded that there was water vapour in the solar atmosphere over spots (Secchi, 1869a).

It was found that disturbances in spots could affect different lines of the same element in different ways. Some iron lines, in particular, could be greatly altered, whilst others retained their normal appearance. In 1870, Young observed the spectrum of a spot which showed the sodium D lines reversed (i.e. with a bright core); he found that he could also see the helium D_3 line in absorption in this spot (Young, 1870). This led to a general interest in the occurrence of line-reversals. It was noted that the calcium H and K lines were always reversed over spots. Sometimes they were doubly reversed (the bright central strip then showing a narrow absorption line down the centre). Young pointed out early on in the development of this work that even when H and K were singly reversed over spots, they were often doubly reversed over the neighbouring faculae. It was found, in comparison, that the D_3 line might sometimes be bright over the umbra of a spot, yet dark over its penumbra.

From time to time suggestions were made that some of the lines in sunspot spectra were spurious. Thus, in the early days, Respighi attributed them to that old standby—instrumental aberrations (see Lockyer, 1874, p. 387). At the end of the century, J. Evershed (1897) suggested much more plausibly that many of the unwidened lines in spot spectra might be due to scattered photospheric light.

For a short time during the middle of the 1860's, there was a vigorous discussion of the physical nature of sunspots. This was sparked off by a theory of the Sun put forward by the French astronomer, Faye (1865). Faye, like most of his contemporaries, believed that the Sun was pervaded by extensive convection currents. He tried to associate this concept with Carrington's measurements of the motions of sunspots, which had been published not long before. Hot gases were supposed to be rising from the solar interior. This retarded the surface rotation of the Sun, but the effect was less at the equator than at higher latitudes because the vertical distance ascended by the currents differed at these places. The gas radiated energy as it rose, cooled, and partially condensed into solid particles (or liquid droplets), thus creating the photosphere. This cooler gas then fell back into the solar interior and was reheated. Faye regarded sunspots as places where the ascending current of gas was particularly strong, so that it blew away the particles forming the photosphere. It therefore became possible to look down into the deep interior of the Sun, where the gases were too hot to emit any visible radiation. As a result, the spot appeared to be dark.

About the same time that Faye's work appeared, the observers associated with Kew—Warren De la Rue, Balfour Stewart and B. Loewy—published a book which discussed the nature of sunspots (for a brief summary, see De la Rue et al., 1865a). Although they based their ideas on the same foundation as Faye—that the Sun was the seat of large-scale convective motions—they came to radically different conclusions. They inferred that spots were associated with descending currents of gas, and were therefore cooler than their surroundings. Again, whereas Faye had supposed the differential rotation of the Sun to result from a lesser retardation at the equator, they attributed it to a greater acceleration there.

From the start, Faye's concept of the solar surface was open to the greater criticism. One objection, emphasized especially by Balfour Stewart, was that his explanation of sunspots ran directly counter to Kirchhoff's relationship for the radiation emitted and absorbed by a hot body. It was argued that, if the gas in the solar interior was indeed, too hot to emit, then it would also be too hot to absorb. Hence, the interior of the Sun should be transparent, and it should be possible to see through a sunspot to the photosphere on the far side. Since this would have the same brightness as the photosphere on the near side, the spot would not appear to be dark at all. It was also urged that visual observations of motions within sunspots accorded better with the idea of flow into a spot rather than out of it; even, perhaps, that spots exerted an actual suction effect on their surroundings. As has been mentioned, however, the decisive test was Lockyer's observation of the sunspot spectrum. It had been reasoned on Faye's theory that the ascending hot gases, as they recombined, would produce an emission spectrum. On the other hand, it was agreed that descending cool gases would produce absorption. The result favouring the latter concept, it became widely accepted throughout the rest of the century, although theories more similar to Faye's continued to be supported from time to time (e.g. by Secchi at the end of the 1860's). In fact, Faye, and the British observers agreed on several points. Thus they both believed that the Sun was predominantly gaseous, that the surface effects were due to convection currents, and that the photosphere was a region in the atmosphere rather than a definite surface. They both agreed, too, on the nature of faculae, which they attributed to hotter, ascending currents. The British observers justified this by their belief that faculae lagged behind spots in the solar rotation: as would be expected if they had been ejected upwards.

The arguments against considering sunspots as regions of ascending hot gas convinced Faye himself and in the early 1870's he proposed an entirely new approach, (Faye 1872). He again explained the differential rotation of the Sun as due to ascending and descending currents of gas. Now, however, he suggested that the relative motions on the Sun's surface led to the formation of eddies. These then developed into full-scale cyclones, which, by analogy with terrestrial cyclones, would be cone-shaped. Material would be sucked into the vortex, thus formed, and carried downwards in a descending current. The overall result would be a sunspot similar in appearance to that suggested by the observers at Kew. Although this theory accorded better with the observations than Faye's earlier theory, there were two major objections which prevented its complete acceptance. In the first place, as has already been mentioned, only a small percentage of all sunspots show a vortical motion. Moreover, those spots which do resemble vortices may show either clockwise or anti-clockwise motion in the same hemisphere, whereas on Faye's theory the motion should always be in the same direction. In the second place (as Young especially emphasized), the relative drift at different solar latitudes was quite small—too small, in fact, to have set up the large-scale vortices attributed to it. Young, himself, preferred to modify an alternative theory proposed by Secchi at about the same time as Faye's second theory. Secchi, like Faye, had abandoned the idea that sunspots were regions of ascending currents in the face of the contrary evidence. Instead, he noted that spots always seemed to be surrounded by faculae and sometimes by prominences. He therefore suggested that faculae and prominences were caused by an eruption around the edge of a ring. The vapours thrown upwards cooled, and fell back into the middle of the ring, forming a central depression. This he identified with the sunspot. (For a comparison of Faye's and Secchi's theories, see: Young, 1896, pp. 182–6.)

Later in the century, J. M. Schaeberle (1890) advocated a theory of sunspot formation rather similar to Secchi's. One difference was the height to which material could be thrown from the solar surface: he believed that it might be ejected as far as the orbits of Jupiter, or Saturn, before falling back again. He also attributed the equatorial acceleration to the effects of the descending material on the Sun's surface. Lockyer (1886) also believed that spots were regions of descending currents of gas. He had been convinced of this even before his spectroscopic work, for he believed he saw signs of such a motion whilst he was observing spots directly in 1865. Surprisingly enough, however, in view of the apparently decisive nature of his spectroscopic observations, Lockyer gradually came round to the view that spots were regions of higher temperature than the surrounding photosphere (this tied in with his dissociation hypothesis). He explained the excess heat as a consequence of the energy released when the downrush of material struck the solar surface.

It was suggested from time to time during the nineteenth century (e.g. by Sir John Herschel) that at least some sunspots were caused by the collision of meteorites with the Sun's surface. The idea was later included by Lockyer as a part of his meteoritic hypothesis (which dealt with the formation and evolution of stars). It was also pursued as a way of explaining the periodicity of the sunspot cycle. One possibility, for example, was that the impacting meteorites (or meteors) followed an elongated path with a period of about 11 years (which means that they would go out about as far as Saturn's orbit). From a comparison with meteor streams striking the Earth, the meteorites would be more densely packed along one part of their orbit than elsewhere. Again, from analogy with the Earth, there would be more than one stream intersecting the Sun's surface. Taken together, these concepts could explain why there were irregularities in the growth and decay of the sunspot cycle (for this type of theory, see: Schuster, 1879).

Another suggestion which received considerable airing was that the occurrence of spots depended on the relative positions of the planets. For example, De la Rue, Stewart and Loewy (1865b) deduced from their solar photographs for the preceding three years, that spots tended to form more frequently when any two of the planets Mercury, Venus and Jupiter were in line. Others suggested that Jupiter alone, or Jupiter and Saturn in conjunction or opposition, were the main influences involved. It was not supposed that the gravitational action of these planets produced sunspots directly, but rather that it acted as some kind of trigger mechanism. The main trouble with these hypotheses was that they seldom fitted the data for more than a few successive years.

Before leaving nineteenth-century theories of sunspots, there is one general point about them that should be made. As we have seen, the predominantly gaseous nature of the Sun, at least in its outer layers, was generally acknowledged by the end of the 1860's. (There were exceptions—Zöllner, for example, still held in the early seventies that the Sun had a liquid surface and that spots were caused by the accumulation of slag-like material.) Despite this, the discussion of solar surface phenomena often continued to regard them as basically comparable with interactions between the solid, or liquid, surface of the Earth and the terrestrial atmosphere. For example, the common picture of a sunspot during the latter part of the nineteenth century depicted it as a shallow depression in the solar surface filled with cool, dense gases. The different types of spectral lines which appeared in spots were due to the coolness of the gases, whilst the odd shapes of the lines were caused by the higher pressure in the depression: the whole concept was visualized implicitly in terms of the terrestrial analogy.

Solar Flares

On 1 September, 1859, Carrington was engaged in his usual task of mapping a sunspot when he suddenly observed the appearance of two bright patches of light. They moved together over the surface of the spot (which was a large one) and finally disappeared five minutes later. Carrington noted that the spot seemed to be precisely the same after their passage as it was before. Similar observations of this event, also from the south of England, were reported by R. Hodgson. These represent the first accounts of a white-light flare in the astronomical literature. (Both papers are reprinted in Part 2.) It was noted by contemporaries that this event took place in the middle of a very intense magnetic storm (it lasted from 28 August to 4 September). Indeed, some accounts emphasized an immediate connection between the flare and its terrestrial effects.

"*At the very instant* of the solar outburst witnessed by Carrington and Hodgson, the photographic apparatus at Kew registered a marked disturbance of all the three magnetic elements" (Clerke, 1885, p. 206).

This was, however, questioned (e.g. by G. M. Whipple, the Superintendent of Kew Observatory), and for many years it remained uncertain whether exactly coincident solar and terrestrial disturbances could occur. Many and varied explanations were proposed for the flare. Generally the analogy with terrestrial aurorae was stressed. Piazzi Smith (1860) did suggest that a pair of large meteorites had hit the solar atmosphere, but he received little support.

Although the white-light flare of 1859 was for a long time regarded as unique, spectroscopic flares were observed quite often from the seventies onwards. An early instance was described by Young.

"On August 3, 1872, the chromosphere in the neighbourhood of a sun-spot, which was just coming into view around the edge of the sun, was greatly disturbed on several occasions during the forenoon. Jets of luminous matter of intense brilliance were projected, and the dark lines of the spectrum were reversed by hundreds for a few minutes at a time. There were three especially notable paroxysms at 8.45, 10.30, and 11.50 a.m. local time. At dinner the photographer of the party, who was determining the magnetic constants of our station, told me, without knowing anything about my observations, that he had been obliged to give up work, his magnet having swung clear off the scale. Two days later the spot had come around the edge of the limb. On the morning of August 5th I began observations at 6.40, and for about an hour witnessed some of the most remarkable phenomena I have ever seen. The hydrogen lines, with many others, were brilliantly reversed in the spectrum of the nucleus, and at one point in the penumbra the C line (of hydrogen) sent out what looked like a blowpipe-jet, projecting toward the upper end of the spectrum, and indicating a motion along the line of sight of about one hundred and twenty miles per second. This motion would die out and be renewed again at intervals of a minute or two. . . . On writing to England, I received from Greenwich and Stonyhurst, through the kindness of Sir G. B. Airy and Rev. S. J. Perry, copies of the photographic magnetic records for these two days. . . . After making allowance for longitude, the magnetic disturbance in England appears strictly simultaneous, so far as can be judged, with the spectroscopic disturbance seen on the Rocky Mountains" (Young, 1896, pp. 166–8).

The relationship between the solar cycle and magnetic disturbances on the Earth was almost universally accepted in the latter half of the century: the last important astronomer to oppose it was Faye, and he acceded to the majority opinion in 1885. However, in 1892 Lord Kelvin, in his presidential address to Royal Society, examined the energy involved in magnetic storms (he used, as an example, one which had been observed in 1885).

"In this eight hours of not very severe magnetic storm as much work must have been done by the sun in sending magnetic waves out in all directions through space as he actually does in four months of his regular heat and light. This result, it seems to me, is absolutely conclusive against the supposition that terrestrial magnetic storms are due to magnetic action of the sun, or to any kind of action taking place within the sun, or in connection with hurricanes in his atmosphere, or anywhere near the sun outside. It seems as if we may also be forced to conclude that the supposed connection between magnetic storms and sunspots is unreal, and that the seeming agreement between the periods has been a mere coincidence" (Thomson, 1892).

The astronomers, although somewhat dismayed by Kelvin's strictures, nevertheless continued to hold to the reality of the relationship. There were a few attempts to consider the phenomena on the Sun and on the Earth as being derived from some third, external cause, but this viewpoint never received very great support.

When the connection between solar and terrestrial activity was first found, it was generally assumed that the causative agent was the sunspots. With the passage of time, however, opinion gradually changed. As the extract from Young (quoted above) shows, the correlation of disturbances on the Earth with sudden disturbances on the Sun was already suspected in the 1870's. As the century progressed, the belief that major terrestrial magnetic fluctuations were specifically connected with chromospheric flares received growing support.

The Temperature of the Solar Surface

Many measurements of the heat radiation from the Sun were made during the latter half of the nineteenth century. A. P. P. Crova (1877), observing from Montpellier, used an adaptation of Pouillet's pyrheliometer; most observers, however, used an actinometer (a thermometer in a constant-temperature enclosure). L. J. G. Violle's instrument, for example, was made up of two concentric metal spheres with a cylindrical hole bored through the middle. A thermometer with a blackened bulb was inserted through the spheres so that it was illuminated by sunlight shining down the hole. Water was circulated continuously between the two spheres, and the solar heat was estimated from the temperature recorded by the thermometer. Violle's measurements with this instrument in 1875 (together with Crova's less accurate results at about the same time) led to an increase in the solar constant from Pouillet's value of $1\frac{3}{4}$ cal cm^{-2} min^{-1} to a new value of about $2\frac{1}{2}$ (Violle, 1877).

In 1880, Langley devised his first bolometer. A thin strip of metal (he ultimately chose platinum) was included into one arm of a Wheatstone bridge, and exposed to sunlight. A similar strip was put in the opposite arm, but was shielded from direct radiation. Variations in resistance were then correlated with differences in the intensity of solar radiation. This instrument was used to study the variation of atmospheric absorption with wavelength. Langley carried out a series of measurements both at sea level and at 15,000 feet in order to test the effect of a known depth of atmosphere. He concluded from his results that previous estimates of atmospheric absorption had been too low; he therefore increased the value of the solar constant to about 3 cal cm^{-2} min^{-1} (see: Young, 1896, p. 298). Langley's work was widely accepted in the astronomical world, although subsequent modifications to the amount of atmospheric absorption were proposed (e.g. by K. Ångström (1890) who allowed for the effects of absorption by carbon dioxide). Measurements of solar radiation were made in Ireland during the 1890's by Wilson and Gray using one of C. V. Boys' radio-micrometers. Their work was thought to be less certain than Langley's owing to the vagaries of the Irish climate, but was important in the determination of the surface temperature of the Sun (see the paper by Wilson reprinted in Part 2).

Several investigations of the heat received from the Sun included a comparison of the amount received from the centre of the disc and from the solar limb. Again, the most significant measurements of this type were made by Langley. He found that near the limb the intensity was reduced to only a half of its central value. His results were confirmed by E. C. Pickering, by Frost and by Wilson. The latter noted that at the extreme limb, the radiation had only 40 per cent of its central intensity. Neither Langley, nor subsequent investigators, could confirm Secchi's observation that the amount of heat radiated depended on the solar latitude (see: Young, 1896, p. 303). This idea therefore disappeared from the scene. An idea which remained throughout the latter years of the nineteenth century was that the solar constant would probably prove to be slightly variable over the solar cycle. It was recognized, however, that other factors—such as a variation in atmospheric absorption—might well mask the intrinsic effect.

As has been remarked earlier, the numerous measurements of the heat radiation from the Sun were partly inspired by the wish to determine the temperature of the solar surface. J. J. Waterston (1860) used Newton's law to derive a solar surface temperature of nearly 13 million degrees Fahrenheit. Secchi (1861), using the same law, suggested an even higher temperature (although he later lowered his estimate appreciably), so did J.-L. Soret. Vicaire (1872), however, estimated from the same data that the surface temperature of the Sun was only 1400 degrees Centigrade—even lower than Pouillet's earlier value. So long as Dulong and Petit's law and Newton's law were both in use (the former mainly by French observers) there remained this dichotomy in the values proposed for the solar surface temperature: the former always gave values of about 2000 degrees Centigrade, the latter gave values of about one million degrees.

It was realized, of course, that the temperature deduced did not correspond to any specific part of the Sun. Rather it represented an integration over the radiating layers. The temperature derived theoretically from the radiation measurements was therefore called the potential temperature or, following Violle, the effective temperature. (Violle used Dulong and Petit's law, but assumed that the Sun radiated less perfectly than a black body. As a result, he ultimately arrived at a rather higher temperature than usual with this law—3000 degrees Centigrade.)

It was realized by the 1870's that neither of the laws for determining the surface temperature could be accepted unreservedly; attempts were therefore made over the next twenty years either to circumvent them, or to replace them. In 1870, Zöllner (1870) made a study of the motions of prominences on the assumption that they were jets of hot compressed gas released from below the solar surface. From their rate of growth he deduced that the temperature just below the photosphere was about 68,000 degrees, whilst the temperature just above was 28,000. (These temperatures were later raised considerably because much higher velocities were subsequently observed in prominences.) Another possible approach was to compare the solar radiation directly with the radiation from the hottest possible terrestrial source. For example, in the early 1870's, Ericsson compared the solar radiation with that from molten iron. A few years later Langley compared the radiation with that emitted by steel in a Bessemer converter. He estimated that, although the steel was at a temperature of over 2000 degrees, the Sun, nevertheless, emitted many times more light and heat (see: Young, 1896, p. 279).

Another possible way of approximating to the solar temperature was via the spectrum of the Sun. Thus H. L. Le Chatelier (1892) compared the solar spectrum in the red with radiation of the same wavelength emitted by heated bodies in the laboratory. He concluded from this that the Sun had an effective temperature of 7600 degrees Centigrade. J. Scheiner (1894) in Germany examined the change in appearance of two magnesium lines in going from an arc spectrum to a spark spectrum. He then compared these lines with their counterparts in the solar spectrum, and so deduced that the region where they were formed on the Sun was at about the same temperature as the electric arc. Besides the possible influence of temperature on individual spectral lines, it was realized that the temperature of a body would also affect the wavelength at which maximum radiation was emitted. Langley had investigated this for the Sun, and had found that the maximum radiation was emitted in the region of the sodium D lines. He pointed out that this wavelength must be related to the surface temperature of the Sun, but as late as the 1890's the exact relationship remained uncertain (see the discussion in Clerke, 1903, p. 66). On the one hand, it was suggested that the wavelength of maximum radiation varied inversely as the square root of the temperature. On the other, the simple inverse proportionality was preferred. Using the former relationship, H. Ebert (1895) deduced a surface temperature of 40,000 degrees Centigrade; using the latter, Paschen (1895) obtained a temperature of somewhat over 5000 degrees.

There was one other obvious approach, and that was to replace both Dulong and Petit's law and Newton's law by some better approximation. Thus Rosetti (1879) suggested from laboratory experiments that the radiation emitted was proportional to the square of the absolute temperature. (This gave an effective temperature for the Sun of about 20,000 degrees.) In the same year, Stefan (1879) proposed that the dependence was as the fourth power of the absolute temperature: a relationship which was subsequently deduced theoretically by Ludwig Boltzmann. Using Stefan's law, Wilson and Gray (1894) obtained an effective temperature for the Sun of about 8700 degrees Centigrade. (Wilson later reduced this to about 6600—see the paper reprinted in Part 2.) Both Wien's law and Stefan's law were regarded with reserve at the end of the nineteenth century. It was thought that they were too simple in form to account for all the complications of the actual situation.

We have noted that the determination of the solar constant involved an estimation of the terrestrial atmospheric absorption. Many investigators also allowed for absorption in the solar atmosphere—as indicated by the decrease in radiation intensity towards the limb. Measurements of the limb-darkening were made not only in the total radiation, but also at a variety of different wavelengths. Langley in 1875 (and Secchi rather earlier) noted that the solar limb was redder, and the centre of the disc bluer, than the total solar radiation. In 1877, Vogel used a special spectral photometer to compare radiation in different wavelength bands coming from different regions of the Sun's surface. He found that in the violet only 10 per cent of the central intensity remained at the extreme limb, whereas in the red the figure was 30 per cent (see: Young, 1896, p. 280). The rapidity with which the absorption increased towards the limb was generally agreed to indicate the presence of a thin, selectively absorbing layer on the Sun. This was, indeed, a main support for the nineteenth-century belief in a "foggy", particulate photosphere.

The infrared line spectrum was also examined in some detail during the last quarter of the nineteenth century. By 1879, Sir William Abney had developed a photographic plate sensitive as far out as 10,750Å. He subsequently measured the positions of 600 absorption lines between this limit and the normal limit of the visual spectrum, but was unable, for the most part, to identify the atoms, or molecules, involved (Abney, 1886). At about the same time, Langley started to examine the infrared line spectrum with his bolometer. This acted as the detecting element in a spectroscope with a rock-salt prism as the dispersive element (gratings were ruled out because the light loss was too great). The mapping was at first done very laboriously by hand, but later on automatic recording was introduced (see the paper by Langley reprinted in Part 2).

At the other end of the spectrum, the growth in knowledge of the ultraviolet line absorption was much more rapid. This was mainly due to the ease of recording the region photographically: it was found to be fairly easy to get down to 3000Å in this way. Indeed, Rowland's atlas of the solar spectrum, which was published at the end of the century, classified absorption lines down to 3000Å in just the same way as lines in the normal visual region.

The Chromosphere, the Prominences and the Corona

Due to a lack of suitable total eclipses, observation of the outer layers of the Sun did not progress very rapidly during the 1850's. Photography was used, however, at a partial eclipse visible in France in 1858 as a means of determining the apparent diameter of the Sun. (By now the daguerreotype had been replaced by the collodion process.) In 1860, another total eclipse visible from Europe occurred. De la Rue transported the photoheliograph from Kew to Spain for this eclipse, and secured three good exposures. Secchi, 250 miles away, also obtained some reasonable images. A comparison of these photographic results suggested strongly that the prominences and the corona were neither ephemeral, nor subjective, but relatively permanent physical objects. It also confirmed the visual observation that the Moon passed in front of the prominences, thus implying their solar origin (see: De la Rue, 1864).

The group of total eclipses which occurred in the late sixties and early seventies were very important in the development of ideas on the solar atmosphere. For that reason they are worth considering in some detail. The first of these, in 1868, was visible from India and Malaya, and was observed by several groups from Britain, France, Germany and Austria. No important photographic advances resulted from this eclipse; several drawings were made, but they were all appreciably subjective. However, major progress was made in spectroscopy of the outer layers of the Sun. It had been agreed during the sixties that one of the major problems to be examined at the next eclipse was the physical nature of the prominences, and that this was best studied spectroscopically. As a result many observers—in particular, Tennant, Janssen, Rayet, Hall, Pogson, Haig and John Herschel (a son of Sir John Herschel) concentrated on this work. Their reports all agreed on the main point: prominences showed bright-line spectra. This meant necessarily that they were gaseous, and, since the hydrogen lines were particularly bright, that they must contain a good deal of this element. The reports differed appreciably, however, in detail. For example, Herschel (1869) noted three bright lines, one of which he identified with the D line in emission, whilst the other two, so he thought, were not coincident with any major absorption line. Tennant (1868), on the other hand, saw five bright lines. Three appeared to coincide with the Fraunhofer C, D and b lines; the other two were near F and G. Rayet (1868) saw nine lines, seven of which he identified as corresponding to the Fraunhofer lines B, D, E, F, G and two at b. The remaining two appeared one between b and F, and the other near G.

These discrepancies were partly due to differences in the power of the instruments used, and partly due to misidentification of lines. Rayet, for example, mistook the C line of hydrogen for the B line (shown at about that time to be caused by terrestrial absorption). More importantly, the yellow emission line seen during the eclipse was generally identified as the Fraunhofer D line.

Janssen, who was observing the chromosphere through a slit parallel to the solar limb, saw two separate spectra, both consisting of five or six bright lines. A glance through the finder telescope revealed that these originated in two distinct prominences. As the brightest lines of both coincided in position with the Fraunhofer C and F lines, he concluded that hydrogen was a major constituent of prominences. He noted, moreover, that the space between the prominences did not produce a spectrum. This appeared to disprove directly Kirchhoff's concept of an extensive, absorbing solar atmosphere. The force of this contention was, however, somewhat offset by the fact that one, or two, other observers at the eclipse did claim to see spectra from the regions between the prominences (probably due to scattered light).

Janssen was much impressed by the brightness of the lines emitted by the prominences. It occurred to him that they might even be visible out of eclipse with the same instrumentation. Clouds came up after the eclipse, but the next morning (19 August, 1868) was fine, and he once more set up his spectroscope with the slit tangential to the solar limb. He immediately saw the same bright lines again, thus confirming his previous conclusion that hydrogen was the most conspicuous element. He found, however, that the yellow line was not precisely coincident with the sodium D line. Besides examining the lines present at a single point, Janssen noted that by slightly altering the position of his telescope he could trace out the whole form of the prominence, and so detect changes in its shape. He found that very considerable changes could occur from one day to the next: a fact which had not previously been proven. Janssen was not alone in noting the brightness of the prominence lines during the eclipse. From their appearance, Herschel also suggested, though more tentatively, that they might be visible out of eclipse.

Janssen wrote a report on his preliminary results and sent it back to the Académie des Sciences in Paris (reprinted in Part 2). Meanwhile, in England, Lockyer, who had not attended the eclipse, was working on precisely the same problem. Some time before, in 1866, Lockyer had decided, after conversations with Balfour Stewart, that prominences were probably clouds of incandescent gas. This suggested to him that they would have a bright-line spectrum, and that they might be visible out of eclipse —given a spectroscope of high enough resolving power. He communicated this idea to the Royal Society (Lockyer, 1866), and then, in 1867, attempted the actual experiment. He found, however, that his instrument had too low a dispersion for the purpose. Huggins (1868a), who made a similar search at about the same time, also reported negative results. Lockyer then commissioned a more powerful instrument. Owing to several delays in construction, the completed spectroscope was not received until October, 1868. By this time, news had been telegraphed by the eclipse observers in India, reporting their success in finding bright lines in prominences. Janssen's fuller report was, however, still *en route*. As soon as his instrument was in reasonable adjustment, Lockyer trained it on the solar limb, and immediately observed the bright lines. He wrote at once to the Royal Society and to the Académie des Sciences reporting his success (the letters are reprinted in Part 2). By a strange coincidence, his communication was received by the latter body at almost precisely the same time as Janssen's: they were both read at the same meeting. Since bright lines were also found to be emitted by the region round the Sun's limb, which Airy had called the sierra, Lockyer deduced that prominences were simply extensions upwards of this envelope. As the predominant emission was due to hydrogen, the envelope had a characteristic red colour. This led to the suggestion that the corresponding layer should be called the chromosphere, a name which Lockyer accepted, and subsequently used. (He was actually not aware originally that the existence of such a continuous layer had been recorded several times previously.)

Once it became certain that the prominence lines could be seen out of eclipse, it was an obvious next step to try and see whole prominences. Janssen's method of moving the entire instrument was very laborious, so other devices—such as a vibrating slit—were considered. Huggins had by now re-entered the field, and he found that, given a sufficiently powerful spectroscope, the whole prominence could be seen simply by widening the slit (Huggins, 1869). (Zöllner (1869) described this method before Huggins, but did not put it into practice until later.) Owing to the insufficient power of his instrument, Huggins was forced to use a red glass filter to eliminate some of the scattered light; Lockyer with his more powerful spectroscope could use a widened slit alone (see the paper reprinted in Part 2). Huggins subsequently experimented with a variety of different filters to try and see the prominences directly without the aid of a spectroscope. In this, however, he was unsuccessful.

As a rule, chromospheric observations out of eclipse were made with the spectroscope slit radial to the solar limb. This made it possible to measure the extent of the emission beyond the limb, and also to compare the wavelength of the emission line with the wavelengths of near-by Fraunhofer lines. Slits tangential to the solar limb were, however, also used on occasion. Secchi (1869), for example, frequently observed the chromosphere successively with a radial and a tangential slit. As a consequence, he was able to note both that the C line of hydrogen was longer in spot latitudes than elsewhere, and also that the line was continuous right round the solar limb.

Besides the spectroscopic studies, an important polarimetric observation concerning the Sun's outer atmosphere was made at the 1868 eclipse. W. R. Campbell (1868) determined the polarization of light from the corona, showing that it indicated the presence in the corona of reflected sunlight. His work was subjected to some criticism: it was pointed out, for example, that reflected sunlight should also show Fraunhofer lines, whereas these remained undetected in coronal light.

At the next eclipse, in 1869, which was visible from North America, several good photographs of the outer layers of the Sun were obtained, but, once again, the major discovery was spectroscopic. W. Harkness declared that he had observed continuous emission from the corona, together with a single bright line in the green region of the spectrum. The existence of this line was confirmed by Young, who measured its position as 1474 on the Kirchhoff scale (see the paper reprinted in Part 2). This position later proved to be incorrect by several Ångströms. At the time, the misidentification caused much perplexity for it was known that there was also a Fraunhofer absorption line due to iron at 1474. This had been seen reversed in the chromosphere, not long before, by both Young and Lockyer, but now, if the new identification was accepted, it was necessary to suppose that the heavy element, iron, actually formed a more extensive envelope round the Sun than the very light element, hydrogen. This was difficult to conceive. To add to the confusion, Young initially identified the 1474 line with an auroral line which had recently been discovered. He also suggested that there were two fainter lines at somewhat greater wavelengths in the coronal spectrum, which behaved rather like 1474, and were therefore related to it.

Ideas on the physical nature of the corona were still at this time marked by their extreme diversity. Harkness reasoned, on the one hand, that the shadow cast at total eclipse by the Moon on the Earth's atmosphere should make the area of sky round the Moon particularly dark. The corona could not, therefore, be attributed to terrestrial atmospheric scattering. On the other hand, Frankland and Lockyer were just publishing the results of their laboratory investigations into the spectra of gases at different temperatures and pressures. These led them to suggest that the pressure at the base of the chromosphere was extremely low (see their paper reprinted in Part 2). (Subsequent work confirmed this: Janssen even claimed that the chromosphere virtually corresponded to a laboratory vacuum.) Hence, it was argued that such a low-density gas would be quite incapable of supporting an extended atmosphere. Lockyer decided that the inner corona—where the prominences were found—probably did have a physical existence, but that the outer corona was most likely to be chromospheric light scattered in our own atmosphere. This suggestion seemed to receive some support at the time from observations which did undoubtedly show such scattering. (One observer reported bright lines everywhere at the 1870 eclipse—even at the centre of the Moon's disc.) It was also pointed out that the 1474 line appeared to be no brighter in the polar rays than in the rifts between them (an observation confirmed by Herschel and Tennant at the 1871 eclipse). This, too, was taken to be evidence for the presence of scattered light; it was suggested, for example, that the rays and the rifts were due to the effect of the serrated lunar limb on the scattering. A further confusion of ideas about the corona was caused after the 1869 eclipse by reports that the coronal light was then unpolarized: polarization in the corona was regarded as one of the best indications of its solar origin. Nevertheless, despite this apparently strong evidence in favour of scattering, Young's alternative hypothesis that the corona was a sort of solar aurora (based on his misidentification of the 1474 line and the terrestrial auroral line) also gained contemporary support.

The question of scattered light came up again at the next total eclipse in 1870 (visible from the Mediterranean area). Several reasonably good photographs of the corona were obtained at this eclipse, particularly by the British observer, Brothers, stationed in Sicily, and the American, Willard, in Spain. It was generally agreed that the photographs of these two observers, when compared, showed a similar disposition of dark rifts and bright rays. This cast some doubt on the validity of light scattering as a mechanism for forming the corona (though it was argued from the opposite side that scattering imposed by the lunar limb would appear similar all along the eclipse path). It also appeared from a comparison of the results of various observers that the outer corona was seen best under good seeing conditions. It was argued that, on the contrary, if the outer corona were due to scattered light, it should appear most extensive when viewed under bad seeing conditions.

The most important discovery at the 1870 eclipse was, however, again spectroscopic, though this time connected with the lower layers of the solar atmosphere. Kirchhoff had argued that the Fraunhofer lines were produced by absorption in the atmosphere above the solar surface. It was realized that, if he was right, then at an eclipse it should be possible to see this part of the atmosphere sticking up beyond the photosphere. Under these conditions an emission spectrum should be produced which would correspond in detail with the normal absorption spectrum. (The normal chromospheric spectrum was obviously not the spectrum sought after for it was in no sense a reversal of the Fraunhofer lines.) Attempts were made at the successive eclipses to try and see this reversed spectrum—there is some suggestion that Secchi may have seen its initial stages of formation in 1868—but the first successful observations were made by Young in 1870 (his description is reprinted in Part 2). Young had looked for signs of emission at the 1869 eclipse, but he had employed a radial slit which, apparently, drowned the emission lines with scattered light. In 1870 he used a tangential slit, and saw the emission lines flashing out briefly before they disappeared again. (They were also observed by Pye, another member of his party.) Because this flash spectrum had such a short duration, Young was unable to make any measurements; he simply recorded his belief that the emission lines represented a straightforward reversal of the Fraunhofer lines. His observations showed that the region of the solar atmosphere producing the lines (the reversing layer) must be of fairly limited vertical extent. This finally disproved Kirchhoff's contention that absorption occurred over a considerable depth of the solar atmosphere. On the other hand, it also tended to disprove a contrary proposal by Faye that all the absorption occurred within the cloud layer of the photosphere itself, in the regions between the clouds. If Faye's idea had been right, there should have been no reversing layer at all, and hence no flash spectrum. In fact, once Young had shown the way, the flash spectrum was observed at virtually every subsequent eclipse. Pogson even saw it at the end of an annular eclipse in 1872.

During the early 1870's, numerous observations of the chromosphere were also being made out of eclipse. In 1872, Young carried out a special investigation into the spectrum of the lower chromosphere as seen over a period of several weeks under good seeing conditions. He detected nearly 300 different bright lines, but found that most of them were only sporadically visible. He was, moreover, puzzled by the selection of lines which appeared.

"The majority of the lines, however, are seen only occasionally, for a few minutes at a time, when the gases and vapors, which generally lie low, mainly in the interstices of the clouds which constitute the photosphere, and below its upper surface, are elevated for the time being by some eruptive action. For the most part, the lines which appear only at such time are simply 'reversals' of the more prominent dark lines of the ordinary solar spectrum. But the selection of the lines seems most capricious; one is taken, and another left, though belonging to the same element, of equal intensity, and close beside the first" (Young, 1896, pp. 205–6).

The 1871 eclipse produced further information both on the corona and on the chromosphere. Janssen claimed that he saw Fraunhofer absorption lines—particularly the D line—in the corona (see the note reprinted in Part 2). This was accepted as supporting the polarimetric evidence that part of the coronal light was due simply to scattered sunlight. Janssen also noted at this eclipse that hydrogen emission could be detected out well beyond the prominences, and was therefore a feature of the inner corona. A similar result was obtained by Lockyer and Respighi (see: Respighi, 1872). Unlike Janssen, these two were both using slitless spectroscopes. (In its simplest form this is just a prism placed before a camera. The atmosphere of the eclipsed Sun then appears as a succession of bright arcs, corresponding to the number of emission lines present.) Respighi observed the resultant spectrum visually through a telescope, but Lockyer used a photographic camera. Several good direct photographs of the corona were also obtained at this eclipse, and were compared with the results from the slitless spectrographs. It was found that the photographic corona and the spectroscopic corona showed very considerable differences. This led to the suggestion, which was entertained briefly, that there were really two different types of corona present.

Five years later, Young made an observation of considerable importance for the development of ideas concerning the corona. He examined at high dispersion the Fraunhofer line which was supposed to correspond to the coronal 1474 line, and found that it could be split into two components. The component to the red he identified as the iron line; the other he attributed to the unknown coronal element (Young, 1896, p. 257). This was, of course, still a misidentification of the line involved, but it considerably eased one of the major contemporary problems: it was no longer necessary to assume that iron vapour extended out further from the Sun than hydrogen. Instead, a new element—coronium —which was lighter than hydrogen could be postulated.

The 1875 eclipse was visible in Asia, and produced the now customary flock of observers. The result, however, were of less significance than those of the previous eclipses. Excellent photographs of the corona were obtained, showing it out to greater distances (up to 30') than had previously been recorded. It was noted, however, that it had changed shape since 1871. At the latter eclipse, Janssen had described the corona as being like the petals of a giant dahlia; now it seemed to have a rectangular outline. The growing suspicions that the corona changed with the sunspot cycle received further support at the next eclipse in 1878 (visible from North America). In this case, the numerous good photographs showed an apparent decrease in coronal luminosity, and characteristic luminous hairs sticking out of the north and south poles of the Sun. The most surprising feature, however, was the remarkable extension of the corona along the ecliptic—it was traced out to a distance of 6° in this direction. This led to the immediate conclusion that the Zodiacal light was simply an extension of the corona; one, more detailed suggestion was that the equatorial bands represented streams of meteors illuminated by sunlight as they fell into the Sun. (For a summary of the observations made at these two eclipses, see: Schuster, 1879.)

It was also found that the bright coronal lines observed at the 1870 and 1871 eclipses were very hard to trace at the 1878 eclipse, (although Young and Eastman observed both hydrogen emission and the 1474 line all the way round the Sun). The reflected sunlight—providing a continuous spectrum with absorption lines—was, however, easily distinguishable. Attempts were also made at this eclipse to measure the thermal radiation from the corona; it was concluded that a significant amount of heat was emitted.

1882 saw another total eclipse, visible this time from Egypt. The Sun was now near maximum activity, and it was observed with satisfaction that the shape of the corona had changed in the predicted manner. It was noted that the coronal light due to reflection was weaker than at the previous eclipse, but the emission lines were stronger. Several new emission lines were distinguished; particularly by Schuster who observed and photographed some thirty of them. This represented the first successful use of an ordinary slit spectrograph in the photography of the emission lines in the inner corona—earlier photographs had all been obtained with a slitless spectograph. The latter type of instrument was, however, still in use: Abney and Schuster (1884) employed it to obtain spectra of the corona in the infrared. It was supposed that the emission observed in this region could be attributed to hydrocarbon molecules.

Several good direct photographs of the corona were also obtained at this eclipse. Some of these were later scrutinized by Huggins, who came to the conclusion that most of the coronal light was concentrated into the violet region of the spectrum. He therefore initiated experiments to see whether he could isolate this part of the spectrum, and so, perhaps, photograph the corona out of eclipse (Huggins, 1883). He first investigated the use of filters, but later decided that he could rely more simply on the properties of the photographic emulsion. The application of this method to solar photography certainly produced corona-like images, and it was thought at the time that Huggins had, indeed, recorded some of the main coronal features out of eclipse. Nor was he alone in this work. Some earlier photographic recording of the corona out of the eclipse had been attempted by Lohse (1883) in Germany; contemporaneously with Huggins, Wright in the United States was experimenting with visual observation of the corona out of eclipse (see: Young, 1896, pp. 266–7). Wright's method was to reflect the solar image into a darkened room, pass it through a violet filter, block out the central photospheric disc, and display the resultant image on a fluorescent screen. He, too, believed that he could distinguish genuine coronal features out of eclipse.

Work at the next eclipse in 1883, visible from the South Pacific, seemed to confirm the validity of these out-of-eclipse observations. Photographs of the corona taken during the eclipse were thought to show certain major features in common with Huggins' photographs taken before and after the eclipse. Many of the other observations made in 1883 simply confirmed earlier results. An important instrumental advance, however, was the first successful photography of the flash spectrum (although only a few of the brighter lines were registered). Other observational advances were indeed claimed, but they seem, perhaps, in retrospect a little less certain. Thus it was claimed that hydrocarbon bands had been seen in emission in a coronal ray. This seemed to confirm that the corona was very much cooler than the photosphere (not that this was generally thought to require confirmation in the nineteenth century—it appeared obvious). C. S. Hastings observed the eclipse with an apparatus designed to juxtapose spectra from opposite limbs of the Sun. He believed that this comparison showed the 1474 line to change as the Moon crossed the solar disc, and argued that this could only be explained by atmospheric scattering. (For a general description of ideas of the solar corona in the mid-1880's, see: Huggins, 1885.) As this all suggests, beliefs concerning the nature of the corona in the 1880's were little less confused than they had been a decade earlier. New theories were still appearing at regular intervals. For example, Huggins (1885) suggested a new variant of the idea that the corona was an electrical phenomenon. He proposed that coronal rays should be considered as analogous with the tails of comets; that is, they were caused by a repulsive force acting outwards from the Sun. Their brightness was due to incessant electrical discharges running along them.

Acceptance of Huggins' and Wright's observations of the corona out of eclipse did not last long. At the succeeding total eclipse of 1886 (visible from the West Indies), Huggins' procedure for photographing the corona was followed during the eclipse itself. The resultant plates showed no trace of any coronal image at all. It thus came to be realized that the coronal forms recorded by Huggins and Wright were, in fact, simply the results of light scattering in the atmosphere of the Earth.

At the end of the 1880's, Tacchini (1889) pointed out that the position of the coronal rays—and, hence, the overall shape of the corona—was linked to the distribution of the prominences round the solar limb. Since this distribution was known to vary during the solar cycle, the connection between the shape of the corona and the activity of the Sun now seemed more easily understandable. Huggins' analogy of polar rays with comets also received some support during the next few years. A. Hansky made a close study of coronal forms during the 1890's and came to the conclusion that the rays were often hollow: a phenomenon which it was thought could also be distinguished in cometary tails (Hansky, 1897).

The end of the century actually saw the disappearance of several misconceptions concerning the properties of the corona. In the first place, A. Fowler obtained some photographs of the coronal spectrum at the total eclipse of 1898 which clearly showed the 1474 line. From an examination of these, Lockyer finally showed that the wavelength of the 1474 line had been wrongly measured: it was, in fact, not connected with the Fraunhofer absorption line which had been assigned to it (Lockyer *et al.*, 1901). By this time, some ten coronal lines had been recognized. It was strongly suspected that they did not all originate in the same element, and that, therefore, there must be some other element present in the corona besides coronium. It was demonstrated also during this period that the hydrogen lines and the H and K lines of calcium, which had previously been observed in the corona, were not physically attributable to it—they were caused by the scattering of the chromospheric emission in the Earth's atmosphere. At the end of the century, C. G. Abbot and C. E. Mendenhall (1900) showed further that the earlier measurements of significant heat emission from the corona were erroneous: their more sensitive instrumentation showed little such effect.

During these last few years of the century, there were further advances in technique, particularly in photography. We have seen that the first photographs of the coronal emission spectrum were made during this period. Similarly the earliest photographs of the coronal reflection spectrum, showing the Fraunhofer lines, date from this time. In 1898, and again in 1900, H. H. Turner (1898) made a photographic study of polarization in the solar corona, taking direct photographs through Iceland spar. Even the difficulty of photographing the inner and the outer corona simultaneously was to some extent circumvented at the turn of the century, for C. Burckhalter (1900) showed how the excess light from the inner corona could be cut down as much as desired by rotating diaphragms.

Several attempts were made to determine the motions of the corona in the latter part of the century. Thus W. W. Campbell noted distortions of the 1474 line, presumably due to Doppler shifts, which indicated the presence of very rapid movements. The major interest, however, was in determining how the corona rotated. Several early observers of the corona had remarked that it appeared to be flickering, as if it were in rapid rotation, but this was quickly recognized to be an illusion (it was, indeed, taken as support for the idea that the corona originated as scattered light in the Earth's atmosphere). The first definite evidence for coronal

E.E.P.—3*

rotation had to wait, however, for measurements of the Doppler shift. This very difficult task was first attempted by H. A. Deslandres (1893). Unfortunately, he chose to measure the K line of calcium which was subsequently shown to be non-coronal in origin. Later he devised a method of more general application whereby the inclination of the emission lines in different parts of the corona could be compared. But the first significant measurements—by Campbell (paper reprinted in Part 2)—showed a most puzzling picture. It seemed that the rotational speed of the corona differed little from that of the photosphere, whereas the chromosphere lying between them rotated faster.

By the end of the century, considerable progress had been made in classifying and cataloguing the chromospheric spectrum. In the mid-1890's, eleven lines were known to be always visible—if the observing conditions were good enough (see: Young, 1896, p. 206). These could all be attributed to hydrogen, helium or calcium, except for one line at 5317Å. This was the wavelength which had previously been associated with the 1474 line of the corona. Although that confusion was to be cleared up, there remained the enigma of this chromospheric line. If the corresponding Fraunhofer absorption line was identified as due to iron, it was difficult to see why this single, unexceptionable line should be chosen to appear from the whole multitude of photospheric iron lines.

Young (1872) recognized early on that the H and K lines of calcium appeared to be in emission in the chromosphere. He was, nevertheless, reluctant to identify them with the corresponding photospheric lines; it seemed unlikely to him that calcium, which was so much heavier than hydrogen, could yet rise to similar heights above the Sun's surface. There was general agreement with his feeling, but it was soon evident that the chromospheric lines were, indeed, due to calcium. Lockyer (1887, p. 240) held that the observation could be explained very easily if it were supposed that the H and K lines were due not to calcium itself, but to some simpler dissociation product. He pointed out that this hypothesis was supported by the appearance of the 4227Å line of normal calcium whose intensity varied in the opposite direction to the H and K lines. Lockyer's argument retained its impact until near the end of the nineteenth century. Sir William and Lady Huggins (1897) then showed that the same increase of H and K relative to 4227Å could be obtained by a reduction in pressure as well as by an increase in temperature. This suggested that the known lower pressure of the chromosphere, as compared with the photosphere, would suffice to explain the different relative intensities of the calcium lines without invoking dissociation. In this event, the abnormal extension of calcium into the chromosphere had to be explained by some unusual property of the calcium atom, not hitherto detected on Earth.

The major interest in the chromosphere from the sixties onward was, however, concentrated on the prominences. The spectroscopic study of the prominences during the last thirty years of the century was especially revealing. Thus it was found that prominence spectra contained lines from elements which were not normally observed in the general chromospheric spectrum, and that the number of lines visible in a prominence depended on the shape and size of the prominence. Photographic work extended these studies into the ultraviolet, and so showed that the line-spectra of prominences were particularly rich in this region. In the 1890's it was found that some of these ultraviolet lines could be identified as coming from krypton and xenon—newly discovered elements which were only then being investigated in terrestrial laboratories.

We have seen that the discovery of the bright-line spectra of prominences immediately led to the conclusion that they were gaseous. From time to time this conclusion was queried. For example, it was observed at some eclipses that a few prominences were white in colour rather than the normal red. It was supposed that the white light was due to incandescent dust particles present in the prominence along with the normal gas. Tacchini, who pressed this particular idea, claimed that the purely spectroscopic study of prominences could not reveal some of the details of their constitution: it would not, for example, show the continuous spectrum of a white-light prominence (see: Pickering, 1886). He provided more direct proof for this viewpoint by comparing drawings made at the 1870 eclipse with simultaneous spectroscopic observations made out of eclipse by himself. These showed, so he believed, that the spectroscopic picture corresponded only to the central core of the prominence seen directly. Despite the evidence he produced, Tacchini's ideas were not widely accepted. Nor were other contemporary suggestions that there were dark prominences on the Sun (due, perhaps, to clouds of dark hydrogen absorbing the light). It was generally believed that any dark patches in the chromosphere were simply holes (see: Wesley, 1897).

The study of prominences out of eclipse was mainly concerned with their shapes and their motions. In 1870, Lockyer, Zöllner and Respighi independently announced that prominences could be divided into two groups (see: Lockyer, 1874, p. 393). These were named (by Lockyer) as eruptive and nebulous. The former were smaller, changed their shape rapidly, and tended to associate with sunspots. The latter were larger, cloud-like, tended to remain unchanged for several days, and did not concentrate in the sunspot zone. As has been mentioned, the two groups also showed different spectral lines. It was natural to question whether such a simple two-fold, division was sufficient. Secchi (1871), in particular, proposed a more complicated classification; the main drawback felt by his contemporaries was that the subgroups were not really distinct—each graded into the next. In the visual region the structure of prominences was mainly examined in the C line of hydrogen (i.e. Ha); the F line was used occasionally, but it was found to give a somewhat less clearly defined image. In the photographic region, the H and K lines of calcium were used, particularly the latter. One of the early puzzles of this work was that the Sun's surface when viewed in this way appeared to be dark. The explanation came when it was realized that these lines were doubly reversed at the base of chromosphere.

Considerable effort was also expended on the study of the distribution of prominences round the solar limb. The use of a widened slit could only reveal prominences one at a time. It was an obvious further step to try and employ a ring slit extending all round the Sun. Such a device was thought of by Zöllner and by Winlock, but was first used effectively by Lockyer and Seabroke (1873). Although positive results were obtained, the ring slit proved to be unsuitable for detailed studies, and was not generally used for gathering data on prominences. The Italian observers, Secchi, Respighi and, later, Tacchini and Riccò, who were especially outstanding in this type of work, used more conventional means (many of their papers can be found in the *Memorie degli Spettroscopisti Italiani*). They found that the number of prominences varied in phase with the solar cycle. Unlike the spots, however, the prominences were not confined only to the lower latitudes; indeed, there appeared to be a secondary maximum in the prominence numbers at about $\pm 75°$. It was noted that prominences seemed to be more closely related to the faculae than to the sunspots. This was accepted as evidence for the belief that faculae and prominences were analogous structures: faculae being derived from eruptions of the photosphere, prominences from eruptions of the chromosphere.

The Italian astronomers investigated not only the structure of prominences, but also that of the chromosphere as a whole. Respighi, for example, examined the serrated edge of the chromospheric disc, and noted that the shapes of the flame-like protrusions varied from day to day. Similar observations were made by Bredichin in Russia and Spörer in Germany. The latter observer noted that the tips of the flames were often inclined in the same direction over considerable stretches of the solar limb. He therefore concluded that there were permanent currents flowing across large areas of the chromosphere (Spörer, 1871). Some years later, however, Tacchini (1876) showed that large-scale inclination of the flames was a feature of solar maximum; it disappeared at solar minimum, and so chromospheric currents, if present, were certainly not permanent. These studies of the chromosphere also showed that its height was not always constant. Trouvelot (1876) claimed to have seen a general decrease in the height of the chromosphere on one occasion. Later, in the eighties, the Hungarian astronomer, J. Fényi (1888), noted that the chromosphere was generally depressed over sunspots. However, all the measurements agreed that at most times the chromosphere had a height of 10–12″ (i.e. 5000–6000 miles). Prominences, on the other hand, frequently reached a height of 1′: in 1880, Young observed one that extended 13′ above the solar surface.

We have seen that prominences were, regarded from the beginning of spectroscopic work, as chromospheric eruptions. Occasional observations contradicted this belief. Thus Secchi noted that some prominences formed as clouds floating above the chromosphere, and were completely unconnected with it (see: Young, 1896, p. 221). It was accepted that prominences of this type must have formed by condensation *in situ*, but no great importance was attached to the observation for such prominences were thought to be very rare—Young recorded that he had only seen three over a period of 20 years. It was, however, the eruptive motions of prominences which most interested observers. Limb observations showed that velocities of over 100 miles per second either towards, or away from, the solar surface were not uncommon. Lockyer suggested that the tangential velocities of these prominences could be obtained at the same time by observing the Doppler shifts present (see the paper reprinted in Part 2). Observations of the C and F lines soon revealed that the tangential velocities were if anything greater than the radial. It should, perhaps, be remarked at this point that the distinction between eruptive prominences and flares was not clearly drawn in the nineteenth century. As a consequence, some of the high velocities observed at the limb were probably due to flares as well as to prominences. But the important result was independent of this confusion: it was calculated that the highest speeds observed actually exceeded the velocity of escape from the Sun, which must therefore be losing material into space. These high velocities in the chromosphere were, indeed, embarrassing from several points of view. R. A. Proctor (1872) pointed out that, despite the low pressures in the solar atmosphere, the resistance to the motion of prominences must be very large. This implied, in turn, that the force required to accelerate the prominences was considerable, and raised the question of its origin. The simplest way out, and the usual one, was to attribute the energy to an unknown source below the solar surface, but some attempts were made to find possible physical causes. It was suggested, for example, that the energy came from the latent heat released when the prominence was formed. Another approach was to deny the reality of the Doppler shifts: perhaps the motions were only apparent, the process involved being a progressive illumination of the solar atmosphere by an electrical discharge. These views, however, were never held by more than a small minority.

Attempts to photograph prominences out of eclipse were made as early as 1870 by Young, but satisfactory results were not obtained until the next decade when dry emulsions replaced the wet collodion process. Then, in 1891, the problem was tackled again by Deslandres in France and Hale in the United States. (Hale's work is discussed in his article on the spectroheliograph reprinted in Part 2. A contemporary assessment of the contributions of both Hale and Deslandres can be found in Winlock, 1892.) Both of them concentrated on an approach discussed previously by Janssen (1869)—the construction of a monochromatic image of the chromosphere using two slits in tandem. The first slit isolated the part of the Sun's surface to be studied; the second slit isolated the spectral line forming the image. By moving the two slits across the photographic plate together, a monochromatic picture of the whole solar chromosphere could be built up. Despite this basic similarity, Deslandres' and Hale's instruments differed in one important respect. Whereas Hale's *spectroheliograph* depended on continuous scanning of a very narrow wavelength band, Deslandres' original *spectro-enregistreur des vitesses* moved discontinuously, but isolated a much wider band of the spectrum. As a result, Deslandres could photograph rapid motions which were Doppler shifted to such an extent as to be completely invisible in Hale's apparatus. On the other hand, he could not duplicate Hale's continuous examination of the solar surface.

There was a certain amount of dispute between the two astronomers over the question of priority in the construction and use of a spectroheliograph. Deslandres (1905) claimed subsequently that he had used not only a *spectro-enregistreur des vitesses*, but also a spectroheliograph from the time he started work on the problem in 1891. He also stated that his analysis of the optical problems had later been accepted by Hale. Hale (1906b) replied that Deslandres never used a continuously moving slit until 1893.

Work in this field had been started originally in order to monitor the prominences all round the solar limb. It was soon found, however, that the instruments recorded features on the disc of the Sun as well. The photographs—usually taken in the K line—had a mottled appearance due to double reversal of the line over large areas of the disc. The bright mottles were found to occur in the same regions that were shown by visual observations to contain faculae. Hale (1893), indeed, at first identified the mottles with the faculae. This led him to discard the traditional picture of faculae as a kind of mountainous elevation of the photosphere, and to regard them instead as cloud-like formations. He speculated further that there was an underlying unity between faculae and prominences. This picture was disputed by Deslandres (1894) amongst others. It was counterclaimed that the mottles were a species intermediate between the faculae and the prominences. Hale subsequently came to accept this second view, and therefore assigned to the mottles a special name—the flocculi. He thought of them as ascending columns of calcium vapour (which he supposed to be less easily condensed than other photospheric gases) and assumed that they were connected with the photospheric granulation (see: Hale and Ellerman, 1910).

In the early years of the present century, Hale, who was by then director of the Yerkes Observatory, attached his spectroheliograph to the new 40-in. telescope there. He found that, besides the mottling, he could also see dark, elongated patches on the disc of the Sun. It subsequently transpired that these dark filaments were simply the projections of prominences crossing the disc. This discovery made it possible to study the development of prominences over much longer periods of time.

Lockyer's Investigations of the Solar Atmosphere

Sir Norman Lockyer was undoubtedly one of the great solar physicists of the last century. His ideas were in many ways seminal, yet, in his own time, he occupied a rather isolated position. He shows in an interesting, if radical way, how ideas on the nature of the Sun changed during the latter half of the nineteenth century. Some of his ideas and observations have been mentioned previously; here we will follow the development of his thought as a whole (for a more detailed discussion, see the articles by H. Dingle in T. M. Lockyer and W. L. Lockyer, 1928).

Lockyer realized from the beginning that an elucidation of the solar spectrum could only rest upon a careful comparative study of laboratory spectra. At the end of 1868, therefore, he teamed up with Edward Frankland, a highly experienced chemist. (The most important papers produced by Lockyer during this period, either individually or with Frankland, are reproduced in Part 2.) Their initial aim was to try and obtain an explanation of the prominence spectra observed by Lockyer. In particular, they were interested in the source of the D_3 line, and in the behaviour of the chromospheric F line (Lockyer had noticed that it broadened out towards the photosphere, and was often displaced relative to the Fraunhofer line). We have seen that the D_3 line was ultimately assigned to a new element—helium. (This was on Lockyer's initiative; Frankland felt at the time that alternative possibilities had not been exhausted.) The shifts of the F line were attributed, again by Lockyer, to the Doppler effect. As the reality of this effect was then regarded by some of his contemporaries as not fully proven, his explanation met with a certain amount of opposition. To add to the difficulty, Lockyer had observed not only shifts, but also major deformations of the line: in one case he saw the F line as apparently double. If these were treated as Doppler effects, too, the velocities indicated were regarded by some astronomers as excessively large. Other observers (e.g.

Respighi) speculated that the effects must be either instrumental, or, at least, imposed by the atmosphere of the Earth.

The explanation of the varying width of the F line was elucidated by the joint laboratory experiments of Frankland and Lockyer. They found that the width of a spectral line depended on the ambient gas pressure. Hence, the widened base of the F line simply indicated that there was an increase in pressure at the bottom of the chromosphere. Although this was reasonable enough in itself, they found further that the chromospheric line-width indicated a much lower atmospheric pressure on the Sun than had hitherto been contemplated. Experiments at higher gas pressures showed that a pressure much above the atmospheric could actually smear out the individual lines into a continuous spectrum.

Frankland and Lockyer were led by their laboratory work into proposing a new model for the structure of the solar atmosphere. They supposed a gaseous photosphere which, being under pressure, emitted a continuous spectrum. The Fraunhofer absorption lines were produced either within this photosphere, or at no great height above it. The photospheric layer was surmounted by the chromosphere which became very tenuous at high altitudes above the surface. The upper region of the chromosphere was composed mainly of hydrogen, although several other elements appeared lower down. Both the photosphere and the chromosphere were subject to continual disturbance: the former layer was occasionally injected into the latter, and the latter sometimes erupted upwards forming a prominence. The low density of the chromosphere in its upper reaches was thought to make it incapable of supporting any further great depth of atmosphere. The corona, therefore, could not be a true physical appendage to the Sun.

Lockyer's ideas are clearly reflected in the instructions issued to British observers at the 1871 eclipse. Those who were making polarimetric measures were told:

"The special object of this observation is to differentiate, if possible, between the corona, on the one hand, and the chromosphere, or whatever else may be self-luminous (be it even a portion of the corona itself), on the other."

The spectroscopic observers were told that their objectives were:

"1. To determine the actual height of the chromosphere as seen with an eclipsed sun; that is, when the atmospheric illumination, the effect of which is doubtless only partially got rid of by the Janssen-Lockyer method, is removed. . . .

2. To determine if there exists cooler hydrogen above and around the vividly incandescent layers and prominences . . ." (Lockyer, 1874, Appendix I).

The results of the eclipse showed, amongst other things, that hydrogen extended out further from the Sun than had previously been supposed. Lockyer now modified his earlier picture by dividing the chromosphere into an inner and an outer part. The inner part was visible in the chromospheric observations out of eclipse; the outer part was seen at eclipses, and included the inner corona which emitted the 1474 line.

Lockyer's view of the solar atmosphere at this point did not differ very greatly from that of many of his contemporaries. He saw it as composed of a succession of shells based on the photosphere. The largest shell contained the very light gas—whatever it was—that produced the 1474 line. The next largest shell consisted of hydrogen; then came shells of calcium, magnesium and sodium. Finally, there was a very complex shell, not extending very far above the photosphere, which contained all the heavier elements. Lockyer later summarised the ideas he held at this time under three headings:

"1. We have terrestrial elements in the sun's atmosphere.

2. They thin out in the order of vapour density, all being represented in the lower strata. . . .

3. In the lower strata we have especially those of higher atomic weight, all together forming a so-called 'reversing layer' by which chiefly the Fraunhofer spectrum is produced" (Lockyer, 1887, p. 303).

Lockyer, like other workers in the field, was struck by the apparent absence of non-metallic elements in the solar spectrum.

He suggested that these elements might be found above the chromosphere where the temperature would be too low for the production of emission lines. On the other hand, they would, if they were there, be able to reflect the sunlight coming from below. This would clear up the discrepancy which then appeared to exist between the distribution of the photographic corona and the spectroscopic corona.

It can be seen that Lockyer's ideas of the solar atmosphere were based upon the same two fundamental assumptions we have mentioned before. It was supposed firstly that the temperature of the solar atmosphere decreased continuously with height, and secondly that the height in the atmosphere reached by a particular element depended on its atomic weight. It was the first of these two assumptions which eventually proved to be the greatest stumbling block in the development of a comprehensive theory of the solar atmosphere. It was the second assumption, however, that caused the greater heart-searching in the nineteenth century. It was, for example, recognized quite early on that the relative heights of calcium, magnesium and sodium, as indicated by the relative lengths of their emission lines above the solar limb, did not agree with the distribution to be expected from their atomic weights.

Lockyer soon noted from his laboratory experiments that the same substance did not always produce precisely the same spectrum (calcium was an obvious example). He also found that the spectrum of an arc source varied according to the part of the arc selected for observation. Some lines (which he distinguished as long lines) could be seen over most of the arc; others (the short lines) were concentrated in the central regions. The difference was obviously related in some way to the variation in temperature from the outer parts of the arc to the centre. Lockyer now made the important discovery that, when an element was only represented by a few lines in the solar spectrum, these always corresponded to the longest lines seen in the arc spectrum.

Kirchhoff and Bunsen (1860) had assumed that the spectrum of an element on the Sun was identical with its laboratory spectrum, and that this latter was in turn, invariable. They did, in fact, notice that the laboratory spectra were not always identical, but the variations were attributed to differences in the amount of the element present. However, Ångström, who subsequently also observed these variations, ascribed them to differences in temperature. Lockyer's observations confirmed this. He also laid to rest another contemporary problem. Astronomers in the early seventies were worried because several elements appeared to be represented in the solar spectrum by some of their spectral lines, but not by others, and this was thought to cast doubt on the validity of their existence on the Sun. Now that Lockyer had shown that this effect was simply a result of temperature differences between the Sun and laboratory sources, the hitherto doubtful elements (such as copper and strontium) could be added to the list of certain solar occurrences.

Lockyer then went on to consider an extension of this work. If the lines emitted by an element varied according to the conditions, would different elements, given the correct circumstances, show some lines in common? This possibility had, in fact, already been suspected. Kirchhoff had noted from his laboratory experiments with Bunsen that some of the bright lines in the spectra of different metals appeared to coincide. Ångström subsequently pointed out a specific instance: some iron lines, so far as could be judged, coincided with calcium lines. (The qualification here is very important: it must be remembered that all these early observations were made at quite low dispersion.) Having convinced himself that lines common to more than one element actually existed, Lockyer turned next to a study of the solar spectrum, to see whether these common lines differed in any way from the other lines present. He concluded eventually that such a difference could be found: the common lines (he called them "basic" lines) were, he thought, also the lines which were broadened in the spectra of sunspots and which appeared in emission in prominences.

Following on these observations, Lockyer's thoughts in the 1870's turned towards the possibility of elements dissociating.

He knew that an increase in temperature could break up molecules into atoms. He therefore postulated that a further increase in temperature could disrupt atoms into something even simpler. Different atoms could produce the same simpler substance on heating, which explained the existence of basic lines common to more than one element. Lockyer therefore related basic lines to the presence of high-temperature break-down products of atoms. Since such lines appeared in prominences, these must be regions of higher temperature. But, equally, he had identified basic lines in spots; therefore these, too, must be at a high temperature—so reversing his previous ideas.

If the temperature varied throughout the solar atmosphere, so must the extent to which atoms were broken down. The solar spectrum thus became a composite of lines formed in different layers of the atmosphere each at its different temperature. Lockyer enumerated his new concepts under three headings, so as to draw the contrast with his earlier ideas:

"1. If the terrestrial elements exist at all in the Sun's atmosphere they are in process of ultimate formation in the cooler parts of it.

2. The Sun's atmosphere is not composed of strata which thin out, all substances being represented at the bottom; but of true strata, like the skins of an onion, each different in composition from the one either above or below. . . .

3. In the lower strata we have not elementary substances of high atomic weight, *but those constituents of the elementary bodies which can resist the greater heat of these regions*" (Lockyer, 1887, pp. 303–4).

Although Lockyer's explanation of the basic lines was more radical than that accepted by most of his contemporaries, he was not alone in considering them to be important. Young, for example, noted that some of the chromospheric lines he observed most frequently were lines common to two, or more, elements.

"Two explanations suggest themselves. The first, which seems rather the more probable is, that the metals operated upon by the observer who mapped their spectra were not absolutely pure—

either the iron contained traces of calcium and titanium, or vice versa. If this supposition is excluded, then we seem to be driven to the conclusion that there is some such similarity between the molecules of the different metals as renders them susceptible of certain synchronous periods of vibrations—a resemblance, as regards the manner in which the molecules are built up out of the constituent atoms, sufficient to establish between them an important physical (and probably chemical) relationship" (Young, 1872).

It will be seen that Lockyer had dropped his previous belief in a very narrow reversing layer where the Fraunhofer lines were formed. Instead, he now held that different lines could be formed at quite different heights. He pointed out that, if this were true, then the flash spectrum seen at eclipses should not, in fact, be an exact reversal of the normal Fraunhofer spectrum. Because the flash spectrum was visible so briefly, this possibility was hard to test, but observations at the 1882 and 1886 eclipses did seem to support his contention that there were differences, although the time was too short to show whether there was exact agreement with his predictions. Lockyer also derived support for his views from the distortions sometimes observed in Fraunhofer lines. He pointed out, in particular, that some iron lines round sunspots were occasionally distorted whilst others were not. Supposing the distortions to be a Doppler effect, the variation could be explained if the iron lines were produced at different temperatures and, therefore, at different heights in the solar atmosphere (where the motions would be dissimilar).

Lockyer now (by 1886) had evolved a new model for the general structure and circulation of the solar atmosphere. He pictured material flowing out of the Sun near the poles (this showed up as the polar rays at eclipses). As the material rose, it lost heat and condensed, falling back to the Sun in middle latitudes. Where the condensed material hit the Sun's surface a spot was produced, whilst the material thrown up by the impact rose into the atmosphere as prominences. This circulation pattern also explained the differential rotation of the Sun. The height from which the material fell back to the solar surface would depend on the distance it had traversed from the poles. At high latitudes the material would fall back from too small a height even to cause a sunspot. Near the equator it would fall back from very great heights, bringing, so Lockyer argued, increased angular momentum, and therefore accelerating the equatorial rotation.

Quite early on in his laboratory work, Lockyer noted the existence of enhanced lines (as he called them). These were lines whose intensity was greater in the spark than in the arc. Initially he paid little attention to them, for he was much more interested in the basic lines. By the 1890's, however, the resolving power of spectroscopes had risen appreciably, and it was becoming apparent that the lines of different elements which had previously been considered as having the same wavelength, actually differed slightly in position. In other words, the basic lines did not really exist. It had also become apparent by this time that an element without altering its physical state could give rise to more than one type of spectrum. Lockyer, therefore, although still holding to the possibility of dissociation, came to regard it in a different light. Instead of the elements breaking down into a few simpler substances, he now thought of them as breaking down into a different form of the same element. In this new view, the enhanced lines played a vital part, for Lockyer assigned them to the dissociated elements. (It should be noted, in passing, that there are occasional ambiguities in Lockyer's later thought on the subject as he sometimes refers to the enhanced lines as basic lines.)

Lockyer claimed support for his new views from the photographs of the flash spectrum obtained at the 1898 eclipse (see the paper by Evershed reprinted in Part 2). These showed that the differences between the Fraunhofer spectrum and the flash spectrum were essentially due to the increased importance of the enhanced lines in the latter. But, once again, correct interpretation was hindered by the assumption that the temperature of the solar atmosphere decreased with height. Since it was obvious that the flash spectrum was at a higher temperature than the Fraunhofer spectrum, this meant that it must be produced at a lower level.

The observations made it certain, however, that the flash spectrum originated in the lower chromosphere. On the other hand, observations by Fowler at the 1893 eclipse had shown that the coronal spectral lines made their appearance directly above the chromosphere. Lockyer was therefore finally forced to suppose that the absorption spectrum was produced in a thin layer between the chromosphere and the corona.

The Structure and Energy Source of the Sun

Starting in 1869, there was an attempt to explain the overall structure of the Sun in terms of a polytropic gas sphere. This was envisaged as a perfect gas of uniform chemical composition in convective equilibrium. Work in this field was initiated by J. Homer Lane (see the paper reprinted in Part 2), and was continued particularly by A. Ritter (1878) and Kelvin (Thomson, 1887) in the nineteenth century, and by R. Emden (1907) in the early years of the twentieth century. Although there was considerable suspicion of the assumption that the gas would be perfect, this model of the Sun fitted in well with the common belief in contraction as the source of the Sun's energy. Despite the objections of the geologists and biologists to the short time-scale it provided, this mechanism for maintaining the heat of the Sun remained popular amongst astronomers throughout the century. Many of the other proposals put forward at this time to explain the supply of energy in the Sun over a longer period were, in fact, made by people from other disciplines. Thus James Croll, a geologist, suggested that the Sun had been formed by the high-velocity collision of two large bodies (Croll, 1889). It was estimated that the resulting supply of energy should last for much longer than the contraction hypothesis allowed. Or, again, the electrical engineer, Sir William Siemens (1881), postulated that the Sun was continually ejecting material at the equator and absorbing it again through the poles. He believed that, if such a circulation existed, the Sun would be able to derive all of its energy from chemical reactions. The gases ejected would already have been burnt; but it was supposed that in space they were broken back to their original form by the action of the Sun's radiation, so that when they entered the Sun once more they were ready for renewed burning. This cyclical process was driven by the rotation of the Sun. Unfortunately, as Kelvin pointed out, the energy of the solar rotation would only suffice to drive the process for slightly more than a hundred years.

CHAPTER III

The New Era In Solar Physics

THE dividing line between early and modern solar physics can be drawn fairly firmly in the second decade of the present century when for the first time there developed a theoretical understanding of the mechanisms involved in the production of atomic spectra. It would, of course, be ridiculous to suppose that this one development provided a solution to all the problems of solar physics, but, equally undoubtedly, progress without it would have been slow, or non-existent. Despite this, the period from 1900 to the beginning of the First World War saw a rapid growth in some fields of solar physics. We will consider in this chapter two important developments during this decade: one theoretical, and one observational. (This means, of course, that much important work is ignored: for example, Evershed's observations (1909) of motions in sunspots.)

The Nature of the Photosphere

On the theoretical side, there was a gradual increase in sophistication in dealing with the photosphere and the solar atmosphere above it. We have seen that the normal picture of the Sun envisaged an opaque photosphere (due to glowing particles) surmounted by a cooler absorbing atmosphere which produced the Fraunhofer lines. Since the depth of the absorbing atmosphere along the line of sight increased towards the limb, this simple model also predicted the existence of limb-darkening. It was found, however, that when the amount of limb-darkening to be expected was worked out in detail, the theoretical result was completely at variance with the well-established observational data. Early in the twentieth century, Schuster (1903) initiated a new investigation into this problem. To try and get better agreement with the observations, he replaced the absorbing atmosphere with an atmosphere which both absorbed and emitted radiation. He assumed that the absorption and emission coefficients were related to each other by some temperature-dependent function. By adjusting the unknown constants present in a suitable manner it proved possible with this type of atmosphere to obtain a good fit for all the limb-darkening observations available. Since the temperature entered into his equations, he was also able to establish a value for the temperature difference between the photosphere and the solar atmosphere above it.

Subsequently, Schuster (1905) attacked the same problem again. This time he considered an atmosphere which not only absorbed and emitted radiation, but also scattered it. He was mainly concerned on this occasion with the formation of absorption and emission lines, but he also necessarily had to consider the continuous radiation flux. Although Schuster, himself, was still thinking in terms of a photosphere consisting of particles, it is evident from his paper that the need for such a photosphere was much less: a solar atmosphere of the type envisaged was capable of producing many of the observed effects by itself.

This point, amongst others, was taken up in a paper published the following year by Karl Schwarzschild (reprinted in Part 2). As we have seen, the entire Sun, including the solar atmosphere, was believed to be in convective equilibrium. Schwarzschild pointed out that the atmosphere might instead be in radiative equilibrium, and that this possibility could be tested by comparing the theoretical values for limb-darkening in the two cases with the observational results. On carrying through the computations he found that the radiative-equilibrium case produced better agreement with the observations. (In distinguishing between radiative and convective equilibrium, Schwarzschild derived the criterion, still in use today, for a gas to be stable against convection.) Indeed, Schwarzschild's explanation of limb-darkening came out to be essentially the same as the modern one: as the limb is approached the observed radiation comes from higher, and therefore cooler layers of the solar atmosphere. This must be contrasted with the normal view at that time, which was that one saw down to the same level—the photosphere—over the whole of the solar disc, but more of the light was absorbed towards the limb. Schwarzchild's calculations indicated not only that the density of the emitting layers was low, but also that it decreased very rapidly with height. This latter result suggested that the sharp limb of the Sun could be explained perfectly well in terms of a purely gaseous atmosphere—a special photospheric layer was unnecessary. Similarly, the photospheric temperature derived by Schwarzchild seemed clearly to be too high for solid particles to endure, whatever their composition. Thus Schwarzschild's paper demonstrated that the concept of a special photospheric layer should be discarded, and to some extent could be without difficulty. In fact, there were immediate objections which prevented this happening. More especially, the apparent clear-cut distinction between the continuous spectrum and the absorption lines seemed to require a similar complete difference between their modes of production. This problem had to wait for its solution on the Bohr theory of the atom.

The Mount Wilson Solar Observatory

On the observational side, the most important development in the early years of the century was the establishment by Hale of the first modern solar observatory on Mount Wilson (see: Adams, 1954). This became immediately not only the main centre for studies of solar spectroscopy, but also the model for later such observatories.

As we have seen, Hale, at the beginning of the century, was the director of the Yerkes Observatory, and was using its 40-in. refractor for spectroheliograph work. He was also at this time experimenting with the use of a coelostat for solar observation. Coelostats had been used before, especially on eclipse expeditions (H. H. Turner, for example, had employed one at both the 1896 and the 1898 eclipses), but they were not in normal use for observations of the Sun out of eclipse. Hale, however, considered it for this purpose, for he was finding that the 40-in. was not entirely suited to the detailed, high-dispersion work that he had in mind. For one thing, he needed a much larger solar image; for another, he required higher dispersion. Any attempts to modify his instrumentation came up against the difficulty that only a limited weight could be attached to the end of a movable telescope. A coelostat with a fixed, horizontal telescope could overcome this problem. Hale and Ritchey therefore set up such a telescope in the grounds at Yerkes. After some delay (due to a fire which destroyed their original equipment) the instrument came into operation in 1903. Although it worked both mechanically and optically as Hale had hoped, results were only fairly satisfactory: partly because the climatic conditions at Yerkes were unsuitable, partly because there were difficulties with air currents round the instrument. (This latter problem was alleviated a few years later, after the move to California, by the introduction of tower telescopes.)

In 1902, Hale was asked to serve on the Advisory Committee on Astronomy of the newly created Carnegie Institution of Washington. It soon became apparent that he might hope to receive a grant from this body to establish a new observatory, so in 1903 he visited California to inspect possible sites. One of these was Mount Wilson: Hale found that he could obtain excellent solar images there, particularly in the early morning. The following year he heard that the Carnegie Institution had, indeed, granted him the money to found the Mount Wilson Solar Observatory. By this time his horizontal telescope (known as the Snow telescope) had already been dispatched to California. It was soon set up, and was brought into operation as a most versatile instrument—it could feed sunlight either into a spectroheliograph, or a large spectrograph, or a bolometer.

Two major types of observation were made in those early days (Hale, 1906a). Twice a day, a spectroheliograph scan was made of the Sun's disc—usually in more than one line. The main work was done in the H line of Ca II and in Hδ, but subsequently redsensitive plates were introduced, and the emphasis shifted rather to Hα. The other main field was high-dispersion spectroscopy of sunspots. Hale was determined to settle the still vexed question as to the nature of the spots. He combined his own solar observations with the laboratory work of King, Adams and Gale and showed conclusively that spots were cooler than their surroundings. This demonstration was clinched in 1906 when his group (and, independently, Fowler in London) found that titanium oxide and other molecules were present in the spectra of sunspots (Fowler, 1906).

We have seen that some of the spectral lines in sunspots were observed quite early on to have a curious spindle-shape. In 1892 Young discovered that in some iron lines this process went further: single lines in the photosphere appeared double in the sunspot. It was generally supposed that the transition from photosphere to sunspot corresponded to a broadening of the absorption line and the superimposition (sometimes) of a central emission component. Hale, however, saw the possibility of another explanation. A study of his spectroheliograph pictures had convinced him that spiral formations round sunspots occurred with significant frequency. He had therefore returned to the old idea that spots were vortices on the surface of the Sun. This led him to the following speculation:

"We know from the investigations of Rowland that the rapid revolution of electrically charged bodies will produce a magnetic field, in which the lines of force are at right angles to the plane of revolution. Corpuscles emitted by the photosphere may perhaps be drawn into vortices, or a preponderance of positive or negative ions may result from some other cause. When observed along the lines of force, many of the lines in the spot spectrum should be double, if they are produced in a strong magnetic field" (Hale, 1908).

Hale invoked here the Zeeman effect which had been discovered in 1896. It was known from laboratory experiments that the line components produced by splitting in a magnetic field were polarized. Hale therefore investigated the double, or spindle-shaped, lines in sunspots for polarization effects. His success in this venture provided good confirmation that the line shapes in spots were due to the intense magnetic fields there. He was then able to go on and deduce from the line splittings the intensities of the magnetic fields (for a summary of this early work on solar magnetism, see the paper by Hale reprinted in Part 2). So began the study of solar magnetism which was to lead eventually to our present understanding of its basic importance in the development of solar activity.

References

ABBOT, C. G. and MENDENHALL, C. E. (1900), *Astrophys. Journ.* **12**, 73.
ABNEY, W. DE W. (1886), *Phil. Trans.* **177**, 457.
ABNEY, W. DE W. and SCHUSTER, A. (1884), *Phil. Trans.* **175**, 267.
ADAMS W. S. (1954), *Pub. Astron. Soc. Pacific* **66**, 267.
AIRY G. B. (1853), *Mem. Roy. Astron. Soc.* **21**, 2.
ÅNGSTRÖM A. J. (1867), *Phil. Mag.* **33**, 76.
ÅNGSTRÖM, A. J. (1869), *Recherches sur le spectre solaire.* Uppsala.
ÅNGSTRÖM, K. (1890), *Wiedemann's Ann.* **39**, 267.
BAILY, F. (1836), *Mem. Roy. Astron. Soc.* **10**, 1.
BECQUEREL, A. E. (1842), *Bibl. Univ. Genève*, **40**, 341.
BREWSTER D. (1834), *Trans. Roy. Soc. Edin.* **12**, 519.
BREWSTER, D. and GLADSTONE, J. H. (1860), *Phil. Trans.* **150**, 157.
BURCKHALTER, C. (1900), *Nature*, **62**, 535.
CAMPBELL, W. R. (1868), *Proc. Roy. Soc.* **17**, 120.
CARRINGTON, R. C. (1863), *Observations of the Spots on the Sun from November 9, 1853, to March 24, 1861, made at Redhill.* Williams and Norgate, London.
CLERKE, A. M. (1885), *A Popular History of Astronomy during the Nineteenth Century.* A and C. Black, London.
CLERKE, A. M. (1903), *Problems in Astrophysics.* A. and C. Black, London.
CORNU, M. A. (1884), *Compt. Rend.* **98**, 169.
CORTIE, A. L. (1890), *Mem. Roy. Astron. Soc.* **50**, 29.
CROLL, J. (1889), *Stellar Evolution, and Its Relations to Geological Time.* Stanford, London.
DAWES, W. R. (1852), *Mem. Roy. Astron. Soc.* **21**, 157.
DAWES, W. R. (1863), *Mon. Not. Roy. Astron. Soc.* **24**, 33.
DE LA RUE, W. (1862), *Phil. Trans.* **152**, 407.
DE LA RUE, W. (1864), *Proc. Roy. Soc.* **13**, 442.
DE LA RUE, W., STEWART, B. and LOEWY, B. (1865a), *Proc. Roy. Soc.* **14**, 37.
DE LA RUE, W., STEWART, B. and LOEWY, B. (1865b), *Proc. Roy. Soc.* **14**, 59.
DESLANDRES, H. A. (1893), *Compt. Rend.* **116**, 1108.
DESLANDRES, H. A. (1894), *Compt. Rend.* **118**, 842.
DESLANDRES, H. A. (1905), *Bull. Astron.* **22**, 332.
DE VAUCOULEURS, G. (1961), *Astronomical Photography.* Faber and Faber, London.
DRAPER, H. (1877), *Amer. Journ. Sci.* **14**, 89.
DRAPER, J. W. (1842), *Phil. Mag.* **21**, 348.
DULONG, P. L. and PETIT, A. T. (1817), *Ann. Chim. Phys.* **7**, 225, 337.
DUNÉR, N. C. (1891), *Recherches sur la rotation du soleil.* Uppsala.
EBERT, H. (1895), *Astrophys. Journ.* **2**, 57.

EGOROFF, N. G. (1883), *Compt. Rend.* **97**, 555.
EMDEN, R. (1907), *Gaskugeln: Andwendungen der Mechanischen Wärmetheorie.* Teubner, Leipzig and Berlin.
EVERSHED, J. (1897), *Astrophys. Journ.* **5**, 248.
EVERSHED, J. (1909), *Mon. Not. Roy. Astron. Soc.* **69**, 454.
FAYE, H. A. E. A. (1865), *Compt. Rend.* **60**, 89, 138.
FAYE, H. A. E. A. (1872), *Compt. Rend.* **75**, 1664.
FÉNYI, J. (1888), *Mem. Spettroscop. Ital.* **18**, 171.
FORBES, J. D. (1836), *Phil. Trans.* **126**, 453.
FORBES, J. D. (1842), *Phil. Trans.* **132**, 273.
FOWLER, A. (1906), *Mon. Not. Roy. Astron. Soc.* **67**, 530.
FRAUNHOFER, J. (1817), *Denk. König. Akad. Wissen. Münch.* **5**, 193.
    (Translated in : J. S. AMES (1898), *Prismatic and Diffraction Spectra.* Harper, New York.)
FROST, E. B. (1896), *Astrophys. Journ.* **4**, 200.
GAUTIER, A. (1852), *Arch. Sci.* **21**, 194.
HALE, G. E. (1893), *Mem. Spettroscop. Ital.* **22**, 198.
HALE, G. E. (1906a), *Astrophys. Journ.* **23**, 1.
HALE, G. E. (1906b), *Astrophys. Journ.* **23**, 92.
HALE, G. E. (1908), *Mount Wilson Solar Observ. Contrib.* No. 26.
HALE, G. E. and ELLERMAN, F. (1910), *Proc. Roy. Soc.* **83**, 177.
HALM, J. (1922), *Mon. Not. Roy. Astron. Soc.* **82**, 479.
HANSKY, A. (1897), *Bull. Acad. St. Pétersbourg*, **6**, 253.
HARKNESS, W. (1867), *Washington Observ.* Appendix II, p. 60.
HASTINGS, C. S. (1881), *Amer. Journ. Sci.* **21**, 41.
HERSCHEL, F. W. (1795), *Phil. Trans.* **85**, 61.
HERSCHEL, F. W. (1800), *Phil. Trans.* **90**, 284.
HERSCHEL, J. (1869), *Proc. Roy. Soc.* **17**, 116
HERSCHEL, J. F. W. (1840), *Phil. Trans.* **130**, 1.
HERSCHEL, J. F. W. (1847), *Results of Astronomical Observations made during the years* 1834–8 *at the Cape of Good Hope.* London.
HERSCHEL, J. F. W. (1864), *Quart. Journ. Sci.* **1**, 222.
HERSCHEL, J. F. W. (1869), *Outlines of Astronomy.* Longmans, Green, London.
HOWLETT, F. (1894), *Mon. Not. Roy. Astron. Soc.* **55**, 73.
HUGGINS, W. (1867), *Mon. Not. Roy. Astron. Soc.* **27**, 131.
HUGGINS, W. (1868a), *Mon. Not. Roy. Astron. Soc.* **28**, 88.
HUGGINS, W. (1868b), *Proc. Roy. Soc.* **16**, 382.
HUGGINS, W. (1869), *Proc. Roy. Soc.* **17**, 302.
HUGGINS, W. (1883), *Proc. Roy. Soc.* **34**, 409.
HUGGINS, W. (1885), *Proc. Roy. Soc.* **39**, 108.
HUGGINS, W. and HUGGINS, M. L. (1897), *Proc. Roy. Soc.* **61**, 433.
JANSSEN, P. J. C. (1865), *Compt. Rend.* **60**, 213.
JANSSEN, P. J. C. (1869), *Report Brit. Ass.* Part II, p. 23.
KIRCHHOFF, G. R. (1861), *Abhand. Berlin Akad.*, p. 63.
KIRCHHOFF, G. R. (1862), *Abhand. Berlin Akad.*, p. 227.
KIRCHHOFF, G. R. and BUNSEN, R. W. (1860), *Phil. Mag.* **20**, 94.
LAMONT, J. (1852), *Ann. Phys.* **84**, 580.
LANGLEY, S. P. (1874), *Amer. Journ. Sci.* **7**, 92.

LANGLEY, S. P. (1876), *Mon. Not. Roy. Astron. Soc.* **37**, 5.
LANGLEY, S. P. (1877), *Amer. Journ. Sci.* **14**, 140.
LE CHATELIER, H. L. (1892), *Compt. Rend.* **114**, 737.
LIVEING, G. D. and DEWAR, J. (1879), *Proc. Roy. Soc.* **28**, 475.
LOCKYER, J. N. (1866), *Proc. Roy. Soc.* **15**, 256.
LOCKYER, J. N. (1874), *Contributions to Solar Physics.* Macmillan, London.
LOCKYER, J. N. (1878), *Proc. Roy. Soc.* **27**, 308.
LOCKYER, J. N. (1886), *Proc. Roy. Soc.* **40**, 347.
LOCKYER, J. N. (1887), *The Chemistry of the Sun.* Macmillan, London.
LOCKYER, J. N., CHISHOLM-BATTEN, Captain, and PEDLER, A. (1901), *Phil. Trans.* **197**, 151.
LOCKYER, J. N. and SEABROKE, G. M. (1873), *Proc. Roy. Soc.* **21**, 105.
LOCKYER, T. M. and LOCKYER W. L. (Eds.) (1928), *Life and Work of Sir Norman Lockyer.* Macmillan, London.
LOHSE, O. (1883), *Astron. Nach.* **104**, No. 2486.
MAJOCCHI, G. A. (1879), *Mem. Roy. Astron. Soc.* **41**, 508.
MAYER, J. R. (1848), *Beiträge zur Dynamik des Himmels in populärer Darstellung.* Heilbronn.
MILLER, W. H. (1833), *Phil. Mag.* **2**, 381.
NASMYTH, J. (1862), *Report Brit. Ass.* Part II p. 16.
OPPOLZER, E. R. VON (1893), *Astron. Astrophys.* **12**, 419, 736.
PASCHEN, L. C. H. F. (1895), *Astrophys. Journ.* **2**, 202.
PETERS, C. H. F. (1855), *Proc. Amer. Ass. Advanc. Sci.* **9**, 94.
PICKERING, W. H. (1886), *Harvard Observ. Ann.* **18**, 99.
PROCTOR, R. A. (1872), *Mon. Not. Roy. Astron. Soc.* **32**, 51.
RAMSAY, W. (1901), *Nature* **65**, 161.
RAYET, G. A. P. (1868), *Compt. Rend.* **67**, 757.
RESPIGHI, L. (1872), *Nature* **5**, 217.
RICCÒ, A. (1897), *Astrophys. Journ.* **6**, 91.
RITTER, A. (1878), *Wiedemann's Ann.* **6**, 135.
RITTER, J. W. (1801), *Gilb. Ann.* **7**, 527.
ROSCOE, H. E. (1906), *The Life and Experiences of Sir Henry Roscoe.* Macmillan, London.
ROSETTI, F. (1879), *Phil. Mag.* **8**, 324, 438, 537.
ROWLAND, H. A. (1889), *Phil. Mag.* **27**, 480.
RUNGE, C. D. T. and PASCHEN, L. C. H. F. (1895), *Nature*, **52**, 520.
RUNGE, C. D. T. and PASCHEN, L. C. H. F. (1897), *Astrophys. Journ.* **4**, 317.
SABINE, E. (1852), *Phil. Trans.* **142**, 103.
SCHAEBERLE, J. M. (1890), *Mon. Not. Roy. Astron. Soc.* **50**, 372.
SCHEINER, J. (1894), *Pub. Astron. Soc. Pacific*, **6**, 172.
SCHUSTER, A. (1879), *Nature*, **19**, 211.
SCHUSTER, A. (1903), *Observatory*, **26**, 379.
SCHUSTER, A. (1905), *Astrophys. Journ.* **21**, 1.
SECCHI, A. (1861), *Nuovo Cim.* **16**, 294.
SECCHI, A. (1864), *Mondes*, **6**, 703.
SECCHI, A. (1866), *Compt. Rend.* **63**, 163.
SECCHI, A. (1868), *Compt. Rend.* **67**, 1018.
SECCHI, A. (1869a), *Compt. Rend.* **68**, 238.

SECCHI, A. (1869b), *Compt. Rend.* **68**, 764.
SECCHI, A. (1871), *Compt. Rend.* **73**, 826.
SIDGREAVES, W. (1895), *Mon. Not. Roy. Astron. Soc.* **55**, 282.
SIEMENS, K. W. (1881), *Proc. Roy. Soc.* **33**, 393.
SMYTH, C. P. (1858), *Phil. Trans.* **148**, 503.
SMYTH, C. P. (1860), *Mon. Not. Roy. Astron. Soc.* **20**, 88.
SORET, J. L. (1867), *Compt. Rend.* **65**, 526.
SPÖRER, G. F. W. (1861), *Astron. Nach.* **55**, No. 1315.
SPÖRER, G. F. W. (1871), *Astron. Nach.* **78**, No. 1854.
STEFAN, J. (1879), *Wien. Bericht.* **79**, 391.
STONEY, G. J. (1867), *Phil. Mag.* **34**, 304.
TACCHINI, P. (1876), *Mem. Spettroscop. Ital.* **5**, 4.
TACCHINI, P. (1889), *Atti Accad. Lincei*, **5**, 763.
TALBOT, W. H. F. (1834), *Phil. Mag.* **4**, 112.
TENNANT, J. F. (1868), *Mem. Roy. Astron. Soc.* **37**, 1.
THOMSON, W. (1887), *Phil. Mag.* **22**, 287.
TROUVELOT, E. L. (1876), *Amer. Journ. Sci.* **11**, 169.
TURNER, H. H. (1898), *Observatory*, **21**, 157.
VICAIRE, J. M. H. E. (1872), *Compt. Rend.* **74**, 31.
VIOLLE, J. L. G. (1877), *Ann. Chim.* **10**, 289.
VIOLLE, J. L. G. (1883), *Compt. Rend.* **96**, 254.
WATERSTON, J. J. (1860), *Phil. Mag.* **23**, 505.
WESLEY, W. H. (1897), *Phil. Trans.* **190**, 202.
WHEATSTONE, C. (1835), *Report Brit. Ass.* Part II, p. 11.
WILSING, J. (1888), *Astron. Nach.* **125**, No. 3000.
WILSON, A. (1774), *Phil. Trans.* **64**, 6.
WILSON, W. E. (1895), *Mon. Not. Roy. Astron. Soc.* **55**, 458.
WILSON, W. E. and GRAY, P. L. (1894), *Phil. Trans.* **185**, 361.
WINLOCK, W. C. (1892), *Smithsonian Report*, p. 709.
WOLF, R. (1852), *Compt. Rend.* **35**, 364.
YOUNG, C. A. (1870), *Journ. Franklin Inst.* **60**, 232.
YOUNG, C. A. (1872), *Nature*, **7**, 17.
YOUNG, C. A. (1876), *Nature*, **15**, 98.
YOUNG, C. A. (1896), *The Sun.* Appleton, New York.
ZÖLLNER, J. C. F. (1869), *Astron. Nach.* **74**, No. 1772.
ZÖLLNER, J. C. F. (1870), *Phil. Mag.* **40**, 313.

Part 2

The papers in this part are printed in the order in which they have been mentioned in Part 1, i.e. the arrangement is not strictly chronological.

1

Solar Observations During 1843†

H. Schwabe

In Dessau

The weather throughout this year was so extremely favourable that I have been able to observe the sun clearly on 312 days; however, I counted only 34 groups of sunspots, most of which were composed of single small spots or points, a few being composed of several large spots. There were three which stood out from all the larger clusters because of their stability. In January, February and March one of these clusters was visible three times, in April, May and June another was visible four times and in July, August and September a third three times. The second of these had the most numerous and largest spots; its largest spot, which was farthest to the West, was recognisable with the naked eye as a tiny dot the first two times it appeared, and on 30th April, the first time it was observed, it measured 1′ 8″·36 at the thickest part and on 31st May, the second time, 1′ 37″·72.

On 149 days, which were spread out fairly equally throughout each month, I noticed no spots and only rarely any significant light-haze; mostly the surface of the sun was uniformly bright and in favourable atmospheric conditions it appeared to be lightly covered with a fine gravel of bright particles.

From my earlier observations, which I have reported every year in this journal, it appears that there is a certain periodicity in the appearance of sunspots and this theory seems more and more probable from the results of this year. Although I described the numbers of groups from 1826 to 1837 in the 15th volume of Astronomische Nachrichten, no. 350, p. 246, I should now like to

† *Astron. Nach.* **20**, No. 495.

add a complete report of all my observations of sunspots up to the present, in which I have indicated the number of days of observation and the days when there were no spots to be seen, as well as the number of groups. The number of groups alone does not give enough information to judge the period as I am convinced that with very great accumulations of sunspots the number of groups is estimated too low, and during a period when there are few spots the estimate of the number of groups is rather too large. In the first instance, several groups often merge together into a single one and in the second, one group may divide into two by the separating off of some spots. For this reason, I am sure the reader will excuse the repetition of facts given in earlier reports.

Year.	No. of Clusters.	Days when no Spots were Observed.	Observation Days.
1826	118	22	277
1827	161	2	273
1828	225	0	282
1829	199	0	244
1830	190	1	217
1831	149	3	239
1832	84	49	270
1833	33	139	267
1834	51	120	273
1835	173	18	244
1836	272	0	200
1837	333	0	168
1838	282	0	202
1839	162	0	205
1840	152	3	263
1841	102	15	283
1842	68	64	307
1843	34	149	324

If one compares the number of groups with the number of days when no spots are visible, one will find that sunspots have a period of about 10 years, and that for five years of this period they appear so frequently that during that time there are very few or no days when no spots at all are visible.

The future will tell whether this period persists, whether the minimum activity of the sun in producing spots lasts one or two years and whether this phenomenon takes longer to build up or longer to decline.

On 10th and 11th April and on 10th May the sun was so particularly clear in a slightly overcast sky that the darkening at its limb showed up very clearly.

Although I kept a careful lookout for the so-called "light-flashes" round the sun I only saw them on 6th May at midday when a great number shot across the field of vision of the telescope, in a direction different from that of the wind and clouds.

2

Observations on the Sun's Store of Force†

H. L. F. HELMHOLTZ

LET us return to the special question which concerns us here: Whence does the sun derive this enormous store of force which it sends out?

On earth the processes of combustion are the most abundant source of heat. Does the sun's heat originate in a process of this kind? To this question we can reply with a complete and decided negative, for we now know that the sun contains the terrestrial elements with which we are acquainted. Let us select from among them the two, which, for the smallest mass, produce the greatest amount of heat when they combine; let us assume that the sun consists of hydrogen and oxygen, mixed in the proportion in which they would unite to form water. The mass of the sun is known, and also the quantity of heat produced by the union of known weights of oxygen and hydrogen. Calculation shows that under the above supposition, the heat resulting from their combustion would be sufficient to keep up the radiation of heat from the sun for 3021 years. That, it is true, is a long time, but even profane history teaches that the sun has lighted and warmed us for 3000 years, and geology puts it beyond doubt that this period must be extended to millions of years.

Known chemical forces are thus so completely inadequate, even on the most favourable assumption, to explain the production of heat which takes place in the sun, that we must quite drop this hypothesis.

† *Popular Scientific Lectures* **2**, 312–15 (trans. E. Atkinson), 1908.

We must seek for forces of far greater magnitude, and these we can only find in cosmical attraction. We have already seen that the comparatively small masses of shooting-stars and meteorites can produce extraordinarily large amounts of heat when their cosmical velocities are arrested by our atmosphere. Now the force which has produced these great velocities is gravitation. We know of this force as one acting on the surface of our planet when it appears as terrestrial gravity. We know that a weight raised from the earth can drive our clocks, and that in like manner the gravity of the water rushing down from the mountains works our mills.

If a weight falls from a height and strikes the ground its mass loses, indeed, the visible motion which it had as a whole—in fact, however, this motion is not lost; it is transferred to the smallest elementary particles of the mass, and this invisible vibration of the molecules is the motion of heat. Visible motion is transformed by impact, into the motion of heat.

That which holds in this respect for gravity, holds also for gravitation. A heavy mass, of whatever kind, which is suspended in space separated from another heavy mass, represents a force capable of work. For both masses attract each other, and, if unrestrained by centrifugal force, they move towards each other under the influence of this attraction; this takes place with ever-increasing velocity; and if this velocity is finally destroyed, whether this be suddenly, by collision, or gradually, by the friction of movable parts, it develops the corresponding quantity of the motion of heat, the amount of which can be calculated from the equivalence, previously established, between heat and mechanical work.

Now we may assume with great probability that very many more meteors fall upon the sun than upon the earth, and with greater velocity, too, and therefore give more heat. Yet the hypothesis, that the entire amount of the sun's heat which is continually lost by radiation, is made up by the fall of meteors, a hypothesis which was propounded by Mayer, and has been favourably adopted by several other physicists, is open, according to Sir W. Thomson's investigations, to objection; for, assuming it to hold, the mass of the sun should increase so rapidly that the consequences would have shown themselves in the accelerated motion of the planets. The entire loss of heat from the sun cannot at all events be produced in this way; at the most a portion, which, however, may not be inconsiderable.

If, now, there is no present manifestation of force sufficient to cover the expenditure of the sun's heat, the sun must originally have had a store of heat which it gradually gives out. But whence this store? We know that the cosmical forces alone could have produced it. And here the hypothesis, previously discussed as to the origin of the sun, comes to our aid. If the mass of the sun had been once diffused in cosmical space, and had then been condensed —that is, had fallen together under the influence of celestial gravity —if then the resultant motion had been destroyed by friction and impact, with the production of heat, the new world produced by such condensation must have acquired a store of heat not only of considerable, but even of colossal, magnitude.

Calculation shows that, assuming the thermal capacity of the sun to be the same as that of water, the temperature might be raised to 28,000,000 degrees, if this quantity of heat could ever have been present in the sun at one time. This cannot be assumed, for such an increase of temperature would offer the greatest hindrance to condensation. It is probable rather that a great part of this heat, which was produced by condensation, began to radiate into space before this condensation was complete. But the heat which the sun could have previously developed by its condensation, would have been sufficient to cover its present expenditure for not less than 22,000,000 of years of the past.

And the sun is by no means so dense as it may become. Spectrum analysis demonstrates the presence of large masses of iron and of other known constituents of the rocks. The pressure which endeavours to condense the interior is about 800 times as great as that in the centre of the earth; and yet the density of the sun, owing probably to its enormous temperature, is less than a quarter of the mean density of the earth.

We may, therefore, assume with great probability that the sun will still continue in its condensation, even if it only attained the density of the earth—though it will probably become far denser in the interior owing to the enormous pressure—this would develop fresh quantities of heat, which would be sufficient to maintain for an additional 17,000,000 of years the same intensity of sunshine as that which is now the source of all terrestrial life.

3

On the Chemical Analysis of the Solar Atmosphere†

G. R. KIRCHHOFF

SINCE I sent in my last report to the Berlin Academy, I have been almost uninterruptedly engaged in following out the investigation in the direction I there indicated. I will not now speak either of the theoretical proof I have given‡ of the facts I there announced, or of the experiments by help of which Bunsen and I§ have shown that the bright bands in the spectrum of a flame serve as the surest indications of the metals present therein; I will take the liberty, in this communication, of informing you of the progress I have made in the chemical analysis of the solar atmosphere.

The sun possesses an incandescent, gaseous atmosphere, which surrounds a solid nucleus having a still higher temperature. If we could see the spectrum of the solar atmosphere, we should see in it the bright bands characteristic of the metals contained in the atmosphere, and from the presence of these lines should infer that of these various metals. The more intense luminosity of the sun's solid body, however, does not permit the spectrum of its atmosphere to appear; it *reverses* it, according to the proposition I have announced; so that instead of the bright lines which the spectrum of the atmosphere by itself would show, dark lines are produced. Thus we do not see the spectrum of the solar atmosphere, but we see a negative image of it. This, however, serves equally well to determine with certainty the presence of those metals which occur

† *Phil. Mag.* **21** (Ser. 4), 185–8, 1861.
‡ *Phil. Mag.* July 1860.
§ Ibid. August 1860.

in the sun's atmosphere. For this purpose we only require to possess an accurate knowledge of the solar spectrum, and of the spectra of the various metals.

I have been fortunate enough to obtain possession of an apparatus from the optical and astronomical manufactory of Steinheil in Munich, which enables me to examine these spectra with a degree of accuracy and purity which has certainly never before been reached. The main part of the instrument consists of four large flint-glass prisms, and two telescopes of the most consummate workmanship. By their aid the solar spectrum is seen to contain thousands of lines; but they differ so remarkably in breadth and tone, and the variety of their grouping is so great, that no difficulty is experienced in recognizing and remembering the various details. I intend to make a map of the sun's spectrum as I see it in my instrument, and I have already accomplished this for the brightest portion of the spectrum—that portion, namely, included between Fraunhofer's lines F and D. By painting the lines of various degrees of shade and of breadth, I have succeeded in producing a drawing which represents the solar spectrum so closely, that, on comparison, one glance suffices to show the corresponding lines.

The apparatus shows the spectrum of an artificial source of light, provided it possess sufficient intensity, with as great a degree of accuracy as the solar spectrum. A common colourless gas-flame in which a metallic salt volatilizes, is in general not sufficiently luminous; but the electric spark gives with great splendour the spectrum of the metal of which the electrodes are composed. A large Ruhmkorff's induction apparatus produces such a rapid succession of sparks, that the spectra of the metals may be thus examined with as great facility as the solar spectrum.

By means of a very simple arrangement, the spectra of two sources of light may be compared. The rays from one source of light can be led through *one* half of the vertical slit, whilst those from another source are led through the *other* half. If this is done, the two spectra are seen directly under one another, separated only by an almost invisible dark line. By this arrangement it is extremely easy to see whether any coincident lines occur in the two spectra.

I have in this way assured myself that all the bright lines characteristic of iron correspond to dark lines in the solar spectrum. In that portion of the solar spectrum which I have examined (between the lines D and F), I have had occasion to remark about seventy particularly brilliant lines as caused by the presence of iron in the solar atmosphere. Ångström only observed three bright lines in this part of the spectrum of the electric spark; Masson noticed only a few more; Van der Willigen says that iron produces only a very few feeble lines in the spectrum of the electric spark. From the number of these lines which I have been able to observe with ease, and map with absolute certainty, some idea may be formed of the capabilities of the instrument which I am fortunate enough to possess.

Iron is remarkable on account of the number of the lines which it causes in the solar spectrum; magnesium is interesting because it produces the group of Fraunhofer's lines which are most readily seen in the sun's spectrum, namely, the group in the green, consisting of three very intense lines to which Fraunhofer gave the name b. Less striking, but still quite distinctly visible, are the dark solar lines coincident with the bright lines of chromium and nickel. The occurrence of these substances in the sun may therefore be regarded as certain. Many metals, however, appear to be absent; for although silver, copper, zinc, aluminium, cobalt, and antimony possess very characteristic spectra, still these do not coincide with any (or at least with any distinct) dark lines of the solar spectrum. I hope before long to be in a position to publish more extended information on this point.

The combination of Ruhmkorff's induction coil with the spectrum apparatus will doubtless also be of importance for the chemistry of terrestrial matter. Very many metallic compounds do not give the spectrum peculiar to the metal when placed in a flame, because they are not sufficiently volatile, but they give it at once when placed on the electrodes of an electric spark. These lines are then indeed seen, together with those of the metal of the electrode, and those of the air through which the spark passes; and owing to the great number of bright lines which compose the spectrum of every electric spark, it would be almost impossible, without a special arrangement, to distinguish the lines caused by the metal of the electrodes from those produced by the metallic salt added. The special arrangement which in this case removes all difficulty, consists in allowing the spark to pass at the same instant between two pairs of electrodes, in such a manner that the light of one spark passes through the upper half of the slit, whilst the light of the other spark passes through the lower half of the slit, so that the two spectra are seen one directly above the other. If both pair of electrodes are pure, both the spectra are alike; if a metallic salt is brought on to one of the electrodes, the lines peculiar to that metal appear in the one spectrum in addition to those present before. These are recognized at the first moment, because they are absent in the other spectrum. The lines which are common to the two spectra may serve, when they are once for all drawn, as the simplest mode by which to represent the position of the lines of the other metals employed.

I have proved that in this way the metals of the rare earths, yttrium, erbium, terbium, etc., may be detected in the most certain and expeditious manner. Hence we may expect that, by help of Ruhmkorff's coil, the spectrum-analytical method may be extended to the detection of the presence of all the metals. I trust that this expectation may be borne out in the continuation of the research which Bunsen and I are jointly carrying on with the object of rendering this method practically applicable.

4

On a New
Proposition In the Theory of Heat†

G. R. KIRCHHOFF

SOME few months ago I communicated to the Society certain observations, which appeared of interest because they give some information respecting the chemical composition of the solar atmosphere, and point the way to further knowledge on this subject. These observations led to the conclusion that a flame whose spectrum consists of bright lines is partially opake for rays of light of the colour of these lines, whilst it is perfectly transparent for all other light. In this statement we find the explanation of Fraunhofer's dark lines in the solar spectrum, and the justification of the conclusions regarding the composition of the sun's atmosphere; for we find that a substance which, when brought into a flame, produces bright lines coincident with the dark lines of the solar spectrum, must be present in the sun's atmosphere. The fact that a flame is partially opake solely for those rays which it emits, was, as I stated at the time, a matter of some surprise to me. Since that time I have arrived, by very simple theoretical considerations, at a proposition from which the above conclusion is immediately derived. As this proposition appears to me to be of considerable importance on other accounts, I beg to lay it before the Society. A hot body emits rays of heat. We feel this very perceptibly near a heated stove. The intensity of the rays of heat which a hot body emits, depends on the nature and on the temperature of the body,

† *Phil. Mag.* **21** (Ser. 4), 241–7, 1861. Abstract of a Lecture delivered before the Natural History Society of Heidelberg. Communicated by Professor Roscoe.

but is quite independent of the nature of the bodies on which the rays fall. We *feel* the rays of heat only in the case of very hot bodies; but they are emitted from a body, whatever be its temperature, although the amount diminishes with the temperature. In proportion as a body radiates, it loses heat, and its temperature must sink unless this loss is made up. A body surrounded by substances of the same temperature undergoes no change of temperature. In this case the loss of heat caused by its own radiation is exactly compensated by the rays which surrounding substances give out, a part of which the body absorbs. The quantity of heat which this body absorbs in a given time must be equal to that which in the same time it emits. This holds good whatever the nature of the body may be; the more rays a body emits, the more of the incident rays must it absorb. The intensity of the rays which a body emits has been called its *power of radiation or emission*; and the number denoting the fraction of the incident rays which is absorbed has been called the *power of absorption.* The larger the power of emission a body possesses, the larger must its power of absorption be.

A somewhat closer consideration shows that the relation between the powers of emission and absorption for *one* temperature is the same for all bodies. This conclusion has been verified in many special cases, both in the last ten years and in former times. The foregoing proposition requires, however, that all the rays of heat under consideration are of one and the same kind; so that these rays are not qualitatively so far different that one part of them are absorbed by the bodies more than another part; for, were this the case, we could not speak of the power of absorption of a body, simply because it would be different for different rays. Now we have long known that there really are different kinds of heating rays, and that in general they are unequally absorbed by bodies. There are both dark- and luminous-heating rays; the former are almost all absorbed by white bodies, whilst the latter rays are thus scarcely absorbed at all. Indeed the variety of the rays of heat is even greater than the variety of the coloured rays of light. The rays of heat, the dark as well as the luminous, are influenced in the same manner as the rays of light, by transmission, reflexion, refraction, double refraction, polarization, interference, and diffraction. In the case of the luminous rays of heat, it is not possible to separate the light from the heat; when one is diminished in a given relation, the other is diminished in the same ratio. This has led to the conclusion that rays of light and heat are essentially of the same nature; that rays of light are simply a particular class of the heat-giving rays. The dark rays of heat are distinguished from the rays of light, just as the differently coloured rays are distinguished from each other, by their period of vibration, wave-length, and refractive index. They are not visible because the media of our eyes are not transparent to them. A difference of quality is noticed amongst the rays of light, not only in respect to the colour, but also in respect to the state of polarization. Hence not only have we to distinguish the heating rays according to the wave-lengths, but we have also to divide rays of one wave-length into those variously polarized. If we take into consideration these various kinds of rays of heat, the conclusions which we had drawn concerning the relation between the powers of absorption and emission cease to be binding.

Whether this relation is still found to exist when these variations are taken into consideration, is a question which has, as yet, not been decided, either by theoretic considerations or by an appeal to experiment. I have succeeded in filling up this gap; and I have found that the proposition concerning the ratio between the power of emission and the power of absorption remains true, however different the rays which the bodies emit may be, as long as the notions of emissive and absorptive powers be confined to *one kind* of ray.

The proposition which I have discovered may be thus more precisely defined: Let a body C be placed behind two screens S_1 and S_2, in which two small openings are made. Through these openings a pencil of rays proceeds from the body C. Of these rays we consider that portion which corresponds to a given wavelength λ, and we divide this into two polarized components, whose planes of polarization, are two planes a and b at right angles to each other, passing through the axis of the pencil of rays. Let the intensity of the polarized component a be E (emissive power). Now suppose that a pencil of rays, having a wave-length $= \lambda$, and polarized in the plane a, falls through the openings 2 and 1 and upon the body C. The fraction of this pencil which is absorbed by the body C is called A (absorptive power).

Then the relation $\dfrac{A}{E}$ is independent of the position, size, and nature of the body C, and is alone determined by the size of the openings 1 and 2, by the wave-length λ, and by the temperature. I will point out the way in which I have proved this proposition. I began by considering that bodies are conceivable which, although very thin, absorb all the rays which fall upon them, or which have the capacity of absorption $= 1$. I call such bodies *perfectly black*, or simply *black*. I first investigated the radiation of such black bodies. Let C be a black body. The body C is supposed to be enclosed in a black envelope, of which the screens S_1 and S_2 are a part, and the two screens are supposed to be connected by a black surface surrounding all. Lastly, let the opening 2 be closed by a black surface, which I will call "surface 2". The whole system is to be considered to possess the same temperature, and to be protected against loss of heat from without by an absolutely non-conducting medium. Under these circumstances the temperature of the body C cannot alter; the sum of the intensities of the rays which it emits must therefore be equal to the sum of the intensities of the rays which it absorbs; and because it absorbs all those that fall upon it, the sum of the intensities of the rays it emits must be equal to the sum of the intensities of the rays which fall upon it. If, now, we suppose the following change: The "surface 2" is removed and replaced by a circular mirror which reflects all the rays falling upon it, and whose centre is in the middle of opening 1. The equilibrium of the heat must still be kept up; the sum of the rays which fall on the body C must still remain equal to the sum of the rays which it emits. But, as it emits just as much as before, the quantity of rays which the mirror reflects upon the body C

must be equal to that which the surface 2 emitted. The mirror produces an image of opening 1, which is coincident with opening 1. For this reason, just those rays come back to the body C, after one reflexion from the mirror, as the body C would have emitted through the openings 1 and 2 if this last one had been open; and the intensity of these rays is equal to the intensity of the rays which the surface 2 sent back through the opening 1. This last intensity however, is, evidently independent of the nature of the body C; and hence it follows that the intensity of the pencil of rays which the body C radiates through the openings 1 and 2, is independent of the form, position, and constitution of the body C; supposing of course that this body is black, and that its temperature is a given one. According to this, however, the qualitative composition of the pencil of rays might become different if the body C were replaced by another black body of the same temperature. This is, however, not the case. If I call e the power of emission of this black body compared with a certain wave-length and a given plane of polarization—that, therefore, which I have called E under the supposition that C is a body of any kind—then e is independent of the nature of the body C, if it only be black. In order to render this evident, a further arrangement is necessary. Into the pencil of rays which passes from the opening 1 towards the surface 2, let us suppose a small plate placed, which is of so slight a thickness that it shows in the visible rays the colours of thin plates; let it be so placed that the pencil of rays is incident at the polarizing angle; let the material of the plate be so chosen that it neither absorbs nor emits a sensible amount of rays; let the envelope joining the screens S_1 and S_2 be so shaped that the image which the plate reflects of the surface 2 lies in the envelope. At the position, and of the size of this image, let an opening in the envelope be made; this I will call "opening 3". Let a screen be so placed that no straight line can be drawn from any point of opening 1 to any point of opening 3 without passing through the screen. Let the opening 3 be now closed with a black surface, which I will call "surface 3". The whole system is then supposed to possess the same temperature; there is therefore in this case equilibrium as regards the heat. This equilibrium is supported by rays which, proceeding from surface 3, suffer reflexion on the plate, pass through opening 1, and fall on the body *C*. These rays are polarized in the plane of incidence of the plate, and contain, according to the thickness of the plate, sometimes more of one, sometimes more of another kind of ray. Let the surface 3 be removed and replaced by a circular mirror whose centre is situated at the spot where the plate reflects an image of the centre of the opening 1; then the rays emitted by surface 3 will no longer fall on body *C*, but instead of them those reflected from the mirror will fall upon it, and the equilibrium of the temperature remains unchanged. If we reflect that it does not matter what thickness the plate possesses, or in what position we turn it round the axis of the pencil determined by passing through openings 1 and 2, we arrive, by means of similar considerations, at the conclusion that the power of emission of the black body *C*, considered with respect to a given wave-length and a given plane of polarization, is quite independent of the constitution of this body. A conclusion which naturally arises from this proposition is, that *all* rays which a black body emits are completely unpolarized.

If we imagine that in the foregoing arrangement the body *C* is not black, but of any other colour, the following equation is found by similar reasoning:

$$\frac{E}{A} = e. \tag{1}$$

This equation indicates that the relation between emission and absorption remains constant for all bodies. The equation may obviously be written

$$E = Ae, \tag{2}$$

or

$$A = \frac{E.}{e} \tag{3}$$

I will now notice some remarkable conclusions derived from my proposition. If we heat any body, a platinum wire for example, gradually more and more, it first emits only dark rays; at the temperature at which it begins to glow, red rays begin to appear; at a certain higher temperature yellow rays are seen; then green rays, until at last it becomes white-hot, i.e. emits all the rays present in solar light. The power of emission (E) of the platinum wire is therefore equal to 0 for the red rays at all temperatures lower than that at which the wire begins to glow; for yellow rays it ceases to be equal to 0 at a rather higher temperature; for green at a still higher temperature, and so on. According to equation (1), the emission-power (e) of a completely black body must cease to be equal to 0 for red, yellow, green, etc. rays at the same temperatures at which the platinum wire began to emit red, yellow, green, etc. rays. Let us now consider the case of any other body which is gradually heated. According to equation (2), this body must begin to give off red, yellow, and green rays at the *same temperatures* as the platinum wire. All bodies must therefore begin to glow at the same temperature, or at the same temperature begin to give off red, and at the same temperature yellow rays, etc. This is the theoretical explanation of an experimental conclusion obtained by Draper thirteen years ago. The intensity of the rays of given colour which a body radiates at a given temperature may, however, be very different,—according to equation (2) it is proportional to the power of absorption (A). The more transparent a body is, the less luminous it appears. This is the reason why gases, in order to glow visibly, need a temperature so much higher than solid or liquid bodies.

A second deduction which I will mention brings me back to my special subject. The spectra of all opake glowing bodies are continuous; they contain neither bright nor dark lines. Hence we can conclude that the spectrum of a glowing *black* body (the term being used in the sense already defined) must also be a continuous one. The spectrum of an incandescent gas consists, at any rate most frequently, of a series of bright lines separated from each other by perfectly dark spaces. If the power of emission of such a gas be represented by $\dfrac{E}{e}$, the relation E has an appreciable value for those rays which correspond to the bright lines of the spectrum of the gas, but it has an inappreciable value for all other rays. According to equation (3), however, this relation is equal to the absorptive power of the incandescent gas. Hence it follows that the spectrum of an incandescent gas *will be* the *converse* of this, as I express it, when it is placed before a source of light of sufficient intensity, which gives a continuous spectrum; i.e. the lines of the gas-spectrum, which before were bright, will be seen as dark lines on a bright ground. A remarkable deduction from my proposition which I will mention is, that, if the more remote source of light is an incandescent solid body, the temperature of this body must be higher than that of the incandescent gas in order that such a conversion of the spectrum may occur.

The sun consists of a luminous nucleus, which would by itself produce a continuous spectrum, and of an incandescent gaseous atmosphere, which by itself would produce a spectrum consisting of an immense number of bright lines characteristic of the numerous substances which it contains. The actual solar spectrum is the converse of this. Were it possible to observe the spectrum belonging to the solar atmosphere with all its attendant bright lines, no one would be surprised to hear that, from the existence of the characteristic bright lines of sodium, potassium, and iron in the solar spectrum, the presence of these bodies in the sun's atmosphere has been ascertained. According to the proposition which I have just laid down, there can, however, be just as little doubt concerning the truth of this assertion, as if we saw the real spectrum of the solar atmosphere.

I will, lastly, mention a phenomenon which, although apparently trivial, was of peculiar interest to me, because I foresaw it theoretically, and afterwards verified it by experiment. According to theory, a body which absorbs more rays polarized in *one* direction than in another, must also emit those rays in the same proportion. A plate of tourmaline cut parallel to the optical axis absorbs, at common temperatures, more of those rays falling perpendicularly, whose plane of polarization is parallel to the axis of the crystal, than of those whose plane is at right angles to the axis. At temperatures above a red heat, tourmaline also possesses this same property, although in a less marked degree. Hence the rays of light which the plate of tourmaline emits perpendicular to its surface must be partially polarized; and, moreover, they must be polarized in a plane perpendicular to the plane of polarization of the rays which have been transmitted by the tourmaline. This theoretical conclusion is borne out by experiment.

5

Summary of Some of the Results obtained at Cocanada, during the Eclipse last August, and Afterwards. A letter from P. J. C. Janssen to the Permanent Secretary

Cocanada, 19th September 1868.

I HAVE just arrived from Guntoor, my observation station for the eclipse, and I am taking advantage of the latest collection of post to send to the Academy† news of the mission which I had the honour to undertake.

I have not enough time to send a detailed account; I hope to do that by the next post. Today I shall merely summarise the main results obtained.

The station at Guntoor has been without doubt a most favourable one; the sky was clear, particularly during the totality, and my powerful telescope of nearly 3 metres focus allowed me to make an analytical study of all the aspects of the eclipse.

Immediately after the totality two magnificent prominences appeared; one of them, more than 3 minutes of arc high, shone with a splendour which is difficult to imagine. The analysis of its light showed me immediately that it was formed by a huge gaseous incandescent column, composed mainly of hydrogen.

The analysis of the circumsolar regions, where Mr. Kirchhoff would place the solar atmosphere, did not give results conforming to the theory of that eminent physicist; it seems to me that these

† The French Academy of Sciences.

results must lead to a knowledge of the true composition of the solar spectrum.

However, the most important result of these observations is the discovery of a method whose principle was conceived during the eclipse itself, and which enables us to study the prominences and circumsolar regions at all times, without having to resort to the interposition of an opaque body in front of the sun. This method is based on the spectral properties of the light from the prominences, light which can be resolved into a small number of highly luminous rays, corresponding to the dark lines of the solar spectrum.

The day after the eclipse this method was successfully applied, and I was able to witness phenomena as if they were caused by a further eclipse which lasted all day. The prominences of the day before were considerably reduced. There was hardly any sign of the large prominence and the distribution of the gaseous matter was quite different.

From that day until 4th October I have constantly studied the sun from this point of view. I have drawn maps of the prominences which show how quickly (often in a few minutes) these huge gaseous masses change shape and position. Finally, during this period, which was like a 17-day eclipse, I collected a considerable number of facts which in themselves indicate the physical composition of the sun.

I am pleased to offer these results to the Academy and to the Bureau des Longitudes in return for the confidence which was placed in me, and to the honour shown in entrusting to me this important mission.

6

Notice of an Observation of the Spectrum of a Solar Prominence†

J. N. Lockyer

October 20, 1868.

Sir—I beg to anticipate a more detailed communication by informing you that, after a number of failures, which made the attempt seem hopeless, I have this morning perfectly succeeded in obtaining and observing part of the spectrum of a solar prominence.

As a result I have established the existence of three bright lines in the following positions:

    I. Absolutely coincident with C.

    II. Nearly coincident with F.

    III. Near D.

The third line (the one near D) is more refrangible than the more refrangible of the two darkest lines by eight or nine degrees of Kirchhoff's scale. I cannot speak with exactness, as this part of the spectrum requires re-mapping.

I have evidence that the prominence was a very fine one.

The instrument employed is the solar spectroscope, the funds for the construction of which were supplied by the Government-Grant Committee. It is to be regretted that its construction has been so long delayed.

I have, etc.,

J. Norman Lockyer.

The Secretary of the Royal Society.

These results were announced to the Paris Academy of Sciences by Mr. Warren De la Rue.

† *Contributions to Solar Physics*, 439–41, Macmillan, 1874.

7

Results of the Spectro-Analytical Observations at Bothkamp Observatory†

H. VOGEL

Bothkamp, August 1871

IN view of the general interest in spectral analysis of heavenly bodies, and being in a position possibly to report something new in this respect, I am prompted to outline the main observations made since the start of this year by Dr. Lohse and myself at the Observatory of Mr. von Bulow. Detailed accounts of these observations are to be published later.

The observations were performed on the large 11-inch equatorial telescope, a fuller description of which appears in *Astr. Nacht.* No. 1843, where the details of the spectral apparatus are also discussed. In regard to the latter, I should like to add that repeated micrometric measurements of about 100 lines of the solar spectrum on the basis of the Ångström Atlas (Spectre normal du Soleil) enabled the screw readings to be expressed directly in wavelengths. Moreover, I should remark that in the observations, the influence of distortions of parts of the spectral apparatus which cannot be easily avoided was minimized so far as possible by measuring the sodium or hydrogen lines after each observation with the instrument in the same position.

† *Astron. Nach.* **78**, No. 1864.

1. Spectral Investigations of the Sun

So much is already known about the interesting forms of the prominences and the rapid changes which take place in them, that I shall confine myself to reporting other observations relating to these gaseous systems. The greatest and highest prominence was observed on 5 March. Its height was 160″ and it looked like a huge fire. Near the solar limb the spectrum of this prominence showed 11 bright lines whose position and height were determined accurately.

	Wavelength in Millionths of a Millimetre	Height of the Lines	
C	656·2	10″	Hydrogen
D_3	587·4	10″	?
	530·7	6″	Iron
b_1	518·3	6″	Magnesium
b_2	517·2	6″	Magnesium
b_3	516·8	5″	Nickel
b_4	516·7	4″	Magnesium
	501·3	4″	Iron
	492·3	6″	Iron
F	486·1	8″	Hydrogen
	434·0	6″	Hydrogen

A cloud which later separated from the main mass of this prominence and which hovered for a long time, had a very intense rotary motion. With a narrow slit the bright prominence line $H\beta$ appeared to be very crooked as if twined round the dark F line of the solar spectrum. The departures on either side of the F line amounted to 0·23 millionths of a millimetre wavelength, which corresponds to a velocity of about 20 miles per second.

On 6 May, Dr. Lohse and I spectroscopically investigated a large sunspot, the nucleus of which was split by two bright bridges of light. When the slit of the spectroscope was set on these bridges, very strong displacements of the spectral lines were seen, in such a way that the parts along the edge of the larger spot rose from the solar surface, whilst those on the edge of the smaller spot came down onto the same. In the centre of the bridge the lines coincided with the lines of the spectrum of the surrounding parts of the solar surface. It was striking that, so far as could be investigated, all the lines in the spectrum were involved in this displacement, in which case there must have been a movement of the light-emitting mass in an arc, i.e. an eruption, and at a velocity of 4 to 5 miles per second, since the displacement of the F line was estimated to be 0·05 millionths millimetres wavelength.

On 9 June, with the aid of a reversing spectroscope made available by Professor F. Zöllner which had been fitted on the large equatorial telescope, we succeeded in demonstrating the rotation of the sun through the displacement of the spectral lines. On 10, 11 and 15 June the observations were repeated using a spectroscope belonging to the equatorial telescope without the reversing principle, and once again we were able to detect clearly the very slight displacement corresponding to the relatively small movement of one point of the solar equator, and Dr. Lohse and I made numerous estimates of the magnitude of this displacement. From the observations, the velocity would seem to be rather greater than has up to now been deduced from the movements of the sunspots. However, in some respects these observations themselves have a considerable degree of uncertainty, whilst in other respects the determination of the wavelength of individual lines in the solar spectrum is not yet so exact for observations of the kind under consideration that one would be justified in drawing any more far-reaching conclusions. Only the conclusion, which is not unimportant for hypotheses about the nature of light, that the motion of a luminous point can lead to a change of wavelength for the light beams emitted from it, can be regarded as certain.

The Zöllner reversing spectroscope had two direct-vision systems with 5 prisms in series, the magnification of the observing system was ×6. The conventional spectroscope consisted of five direct-vision prisms and five very high dispersion prisms in a circle. The magnification of the observing system was ×9, but ×24 on 15 June. In the observations with the reversing spectroscope with the slit tangent to the solar limb, the lines of the superimposed spectra were so placed that every one fell exactly on the extension of another. If the slit was tangent to the other solar limb, without any change being made in the apparatus, the exact superimposition was lost and the lines became displaced relative to each other. In the observations with the other spectral apparatus, if the cross-hairs at the focal point of the telescope were adjusted to one line of the spectrum when the slit was on one solar limb, there was no coincidence in respect of the other solar limb. In order to prevent any possible changes taking place, the precaution was taken of fixing the telescope and bringing the image of the sun past the slit during the daily motion. The exposures were therefore always made on the leading solar limb.

8

On a new Method of
Observing Contacts at the Sun's
Limb, and Other Spectroscopic
Observations during the Recent Eclipse†

C. A. YOUNG

OUR party was one of those organized, and subsidized by Prof. J. H. C. Coffin, Superintendent of the Nautical Almanac; and the one which was so fortunate as to enjoy his personal presence and supervision.

Our station was at Burlington, Iowa, a few miles north of the central line, where the duration of the totality was 2 minutes and 50 seconds. The air was perfectly clear, without cloud or haze; the thermometer during the Eclipse ranged from 76° to 67°, and the barometer from 30ᵐ 140 to 30ᵐ 065. The wind was from the NW., but very gentle.

The spectroscopic combination employed, was compiled for the occasion from various instruments belonging to Dartmouth College, and differed so much in the relative proportion and arrangement of its parts from those hitherto used, that a brief description is perhaps necessary.

The telescope which formed the solar image was a comet-seeker by Merz & Son, of 4 inches aperture and 30 inches focal length. An ordinary Huyghenian eye-piece enlarged the image so that when it fell upon the slit of the spectroscope at a distance of 5 inches, it was $2\frac{1}{8}$ inches in diameter. The use of an eye-piece gave an easy means for securing the accurate focus of the limb at the slit, an adjustment of great importance. The spectroscope

† *Amer. Journ. Sciences and Arts* (2nd series), **48**, 370–8, 1869.

proper had telescopes of $2\frac{1}{4}$ inches aperture and $16\frac{1}{2}$ focal length (by Alvan Clark). The eye telescope was provided with an eyepiece magnifying 18 times, and a wire micrometer, constructed from a reading microscope, for determining the position of any new lines in the spectrum by referring them to those already known. This, although a very accurate method, was too slow to be well adapted to eclipse observations, but was the only arrangement I could construct with the time and means at my command.

The collimator had a slit $\frac{1}{8}$ inch long and of adjustable width. It was provided with a small prism, which could be turned up so as to throw into half the slit light from an electric spark formed between platinum electrodes by a small induction coil and Leyden jar.

It also carried a thin brass disc about $2\frac{1}{2}$ inches in diameter, placed in front of the slit, with a hole of $\frac{1}{8}$ inch in the centre. This disc was covered with white paper and graduated into sectors of $10°$ by lines radiating from the centre. This graduated screen, upon which the image of the sun was clearly visible even during the totality, answered the purpose of a finder, and its graduation furnished the means of determining within less than $3°$ the position of any object observed on the sun's limb, or of bringing any desired portion of the limb to the slit.

The spectrum was formed by a train of 5 prisms of $45°$ each, with faces $2\frac{1}{4}$ by $3\frac{1}{4}$ inches. They gave a dispersion of about $18°$ between A and H, with a total deviation of about $165°$ for the D line. The box which contained them was so connected by a link with the arm which carried the eye-telescope, that whenever the latter was moved by its tangent-screw along the spectrum the prism box would turn through an angle just half as great. Thus the prisms were kept in the position of best definition for whatever lines were in the middle of the field of view, the extent of which was sufficient to embrace D and E together.

The telescope and spectroscope proper were firmly secured to a wooden framework, and this was mounted equatorially, with slow motion screws in both right ascension and declination. Fig. 1 gives an idea of the appearance of the whole.

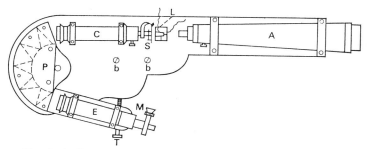

FIG. 1. A, Comet-seeker; C, collimator of spectroscope; L, camera lucida prism, and platinum electrodes; S, graduated disc immediately in front of slit; P, prism box; E, eye telescope; T, tangent screw for moving telescope across the spectrum; M, micrometer; b, b', bolts securing the frame-work to the equatorial mounting.

The spectrum was about $1\frac{3}{4}$ inches broad (referred to a distance of 10 inches) and about 45 long. It showed all the lines on Kirchhoff's maps of the spectrum; such lines as the nickel line between D_1 and D_2 being perfectly distinct.

I had seen with the instrument before the eclipse the following six bright lines at the sun's limb, viz. C, 1017·5 K (near D); 1474 K (near E); F; 2796 K (near G); and h: on two occasions also I had seen a 7th line at† or near 660 K a little below C. The line 1474 K was discovered by me on the 9th of July, and I do not *know* that it was seen by any one else before the eclipse, though quite possibly it is the new line reported by Rayet on the 7th of June, the position of which I have never seen stated. P.S. It was discovered by Lockyer on June 6th.

The scale of my instrument was referred to that of Kirchhoff by repeated and closely accordant measures of 42 intervals between important lines, covering the whole length of the spectrum from A to G. Above G, Ångström's map was made the basis.

---

† Possibly this may be the "fainter red flash" already described by Mr. Lockyer, to which Lieut. Herschel alludes in a letter to Mr. Huggins published in the September number of the *Chemical News*, p. 152, American edition. I have not yet seen Mr. Lockyer's notice.

In order to observe the first contact I had provided myself with a solar eye-piece which could be used upon the comet-seeker without disturbing the spectroscope; but about half an hour before the totality the idea presented itself that a more accurate result could be obtained by means of the spectroscope than in any other way—thus:

Suppose that by means of the tangent screws and the position lines on the screen-disc, that part of the sun's limb at which the contact is to take place is brought across the centre of the slit and kept there by the slow motion.

Under ordinary circumstances, if the eye-telescope is directed upon the C line, the appearance will be what I have roughly

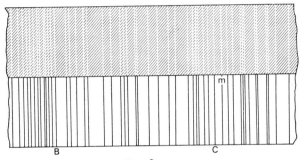

FIG. 2.

represented in Fig. 2. The spectrum will be divided into two portions by a longitudinal line of demarcation. One half of it, formed by the disc of the sun itself, will be intensely bright; the other half formed mainly by the illuminated air near the sun's limb, comparatively dark. Most of the spectral lines will extend across both portions alike; not so the C line. *A brilliant needle of scarlet light* projects into the dusky spectrum from the end of the *dark line* in the bright portion of the spectrum (at the point marked *m,*) formed by the light from the hydrogen in the solar chromosphere. The length of this needle in my instrument usually ranges from $\frac{1}{8}$ to $\frac{1}{4}$ inch, even when the slit is not near any prominence. (The whole width of the spectrum, $1\frac{3}{4}$ inches, was found, with the eye-piece used on the comet-seeker, equivalent to 1′ 55″ of arc, whence the thickness of the chromosphere is easily deduced; seems to vary from 3000 to 8000 miles commonly.)

Now it is evident that as the moon approaches the sun she must first eclipse this chromosphere, and the observer can determine the instant of contact by noticing the extinction of the bright needle.

Similar remarks apply to F, or any other line of the chromosphere spectrum, but C is incomparably the best to observe.

Having arranged my instrument therefore, with the computed point of contact across the center of the slit, I had the unspeakable gratification of seeing everything take place as expected. First, a full half minute before the time of contact, the sharp point of the needle was truncated by the dark edge of the moon, then it grew steadily shorter, (not *less brilliant* what remained of it) until finally its last spark vanished, the C line became exactly like its neighbors, and the contact was effected.

The observation was as easy and definite as that of the transit of a moderately slow star, β Ursæ Minoris for instance. I am confident the observation may be relied upon within a fraction of a second, although it was from 5 to 15 seconds earlier than the time assigned by any of the other observers. I am informed by Professor Mayer, however, who had charge of the photographic operations of our party that it agrees within one third of a second with the time deduced from a preliminary measurement of a photograph taken about 15 seconds after the contact was announced.

With an instrument of sufficient dispersive power, the slit might be opened somewhat widely, and placed *tangent* to the sun's limb. In this case a slight error in the estimated point of contact would not interfere with the accuracy of the observation.

I wish to call attention to the applicability of this method at the coming transits of Venus. It is not possible, perhaps, to predict just how great will be the effect of her atmosphere; but it is difficult to see in what respects this method will suffer from it more than any other. It certainly presents this great advantage, that the observer will perceive and watch the planet's approach long before the instant to be observed, and thus have all the benefit of preparation.

It seems likely also that the instant of *internal contact* will be more easily seized with the spectroscope than with any other instrument. Instead of the rupture of a black ligament, it ought to show the sudden formation of a brilliant line running the whole length of a before dusky spectrum, a phenomenon much more striking than the other.

While the moon was advancing upon the sun, special attention was paid to the appearance of the spectrum lines near her limb. They came up to the edge perfectly square and straight, even when the limb made an angle of only 5° or 6° with the slit; and the longitudinal line of demarcation, before referred to between the brilliant and dusky portions of the spectrum was hard and sharp, in striking contrast with the effect of the sun's limb, which under similar circumstances always gives a boundary more or less hazy and indefinite, and this to a degree continually changing from minute to minute. This contrast was beautifully exhibited a few seconds before the totality when the limbs of both sun and moon were on the spectrum together, the width of the visible portion of the sun having become less than the length of the slit. It was at first thought that this appearance was decisive against the existence of a lunar atmosphere however rare; but a little consideration shows that on the other hand it is if anything, favorable, being a simple consequence of that brightening of the sun's disc near the moon's limb, which is so beautifully evident upon the photographs; and which is most easily accounted for by admitting a slight refraction suffered by that portion of the sunlight which grazes the moon. Possibly, however, it may yet be explained as a case of simple inflection of light.

Before the eclipse began, the existence of prominences on the limb of the sun had been ascertained in the following positions, (reckoning from the north point through the east). A large but faint one near +90°, a small but bright one at +146° (the photographs show *two* here), a long low one at −70°, very near the point of first contact, and an enormous and very bright one at $-130°$, with several others of small elevation but considerable length on different parts of the limb.

At the beginning of the totality the slit was upon the prominences at $+146°$; the eye telescope upon C. As soon as totality commenced this line blazed out magnificently, but from the small extent of the prominence did not reach across the spectrum. No line appeared below C, nor any between C and D. The orange line which for convenience I will call D_3 (1017·5 K) was beautifully bright, but no longer than C. Between it and the next prominent line were two faint lines, situated by estimations at 1250 ± 20 and 1350 ± 20 of Kirchhoff's scale. Then came the 1474 K line, which was very bright, though by no means equal to C and D_3; but attention was immediately arrested by the fact that, unlike them, it extended clear across the spectrum, and on moving the slit away from the protuberance it persisted, while D_3, visible in the edge of the field, disappeared. Thus it was evident that† *this line belonged, not to the spectrum of the protuberance, but to that of the corona.* My impression, but I do not feel at all sure of it, is that the two faint lines between it and D_3, behaved in the same manner, and are also corona lines.‡

I may as well confess that my uncertain memory here is due to the fact that just at this time, while my assistant was handing me the lantern with which to read the micrometer head, I looked over my shoulder for an instant, and behind the most beautiful

† On two or three occasions previously I had been very much surprised at not being able to detect this line in the spectrum of unusually bright prominences. On the other hand I once found it very easy to see at a place on the sun's limb where the other chromosphere lines, usually far more brilliant, were almost invisible.

‡ A careful examination of the photographs, especially No. 2 of the Burlington totality pictures, somewhat diminishes my confidence in the conclusion of the text as to the nature of these three lines (1250, 1350 and 1474). They certainly do not belong to the spectrum of the *most brilliant portion* of the prominences; but around the prominences of the eastern limb, on which the slit of the spectroscope was directed during the first half of the totality, the photograph shows a pretty extensive and well defined nebulosity, evidently distinct from, though associated with, the brilliant nuclei. Now it is *possible* that these lines may belong to this nebulosity, and not to the corona proper;

and impressive spectacle upon which my eyes have ever rested. It could not have been for 5 seconds, but the effect was so overwhelming as to drive away all certain recollection of what I had just seen. What I have recorded I recall from my notes taken down by my assistant.

By this time the moon had advanced so far that it became necessary to shift the slit to the great prominence of the opposite side of the sun. While my assistant was doing this I suppose I must, in the excitement of the moment, have run my eye-piece over the region of the magnesium lines (b), and thrown them out of the field before he had brought any thing upon the slit. At any rate I saw nothing of these lines, which were evident enough to several other observers, and can think of no other way to account for their having escaped me.

The F line in the spectrum of the great protuberance was absolutely glorious, broad at the base and tapering upward, *crookedly*, as Lockyer has before often observed. Next appeared a new line,† about as bright as 1474, at 2602 ± 2 of Kirchhoff's scale. Its position was carefully determined by micrometrical reference to the next line, 2796 K (Hydrogen 7) which was very bright—h was also seen, very clear but hardly brilliant. In all I saw 9 bright lines.

A faint continuous spectrum, without any traces of dark lines in it, was also visible, evidently due to the corona. Its light, tested by a tourmaline applied next the eye, proved to be very

---

for I cannot recall with certainty whether 1474 retained its brilliance at any considerable distance from the prominences or only in their immediate neighbourhood. My strong impression however is that the former was the case, and that the text is correct. I am confirmed in this opinion by Professor Pickering's observation. He used a single-prism spectroscope, with the slit of the collimator simply directed to the sun, and having no lens in front of it. With this arrangement he saw only 3 or 4 bright lines, *the brightest near E* (1474). Now this is exactly what ought to occur, if that line really belongs to the corona, which, from its great extent, furnished to his instrument a far greater quantity of light than the prominences.

† This is undoubtedly the line described by Lieut. Herschel as between F and G, in the letter referred to in a former note, *Chem. News*, Sept., 1869, p. 152, Am. Ed.

strongly polarized in a plane passing through the centre of the sun. I am not sure, however, but that this polarization, as suggested by Prof. Pickering, may have been produced by the successive refractions through the prisms. This explanation at once removes

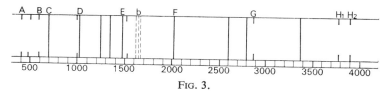

FIG. 3.

the difficulty otherwise arising from the absence of dark lines. Figure 3 exhibits the position of the lines seen by me, and I have also added the magnesium lines.

My observations decide nothing as to specific differences between the different protuberances, since from the smallness of my field of view, I was obliged to observe a portion of the spectrum on one of the prominences and the rest on another.

I had just completed the measurement of 2602 when the totality ended. This line disappeared instantly, but 2796 was nearly a minute in resuming its usual faintness. I cannot describe the sensation of surprise and chagrin, of wasted opportunity, personal imbecility, and complete exhaustion which overwhelmed me when the sunlight burst out. Many other observers I believe shared in it.

Having for some time entertained the idea that there might possibly be some connection between the Aurora Borealis and the Solar Corona, immediately after the eclipse I examined Ångström's map, in which he lays down one bright line in the Aurora spectrum. It plainly did not coincide with 1474. But on my return home I found in the July number of this Journal the positions of five lines observed by Professor Winlock in the spectrum of the Aurora. In numbers of Huggins's scale they are given as follows, viz. 1280 (brightest), 1400, 1550, 1680 and 2640. Now from the data given in Professor Gibbs's recent articles on "wave lengths by the method of comparison," we easily find,

with a probable error of not more than 2 or 3 divisions, that
H 1280 = K 1247; H 1400 = K 1351; and H 1550 = K 1473. I
observed in the corona spectrum K 1250, K 1350 and K 1474.
These coincidences are certainly very remarkable, especially the
last. The positions of the faint lines (K 1250 and 1350) being only
estimations, very little weight can of course be given them. But
as to the position of 1474 in the corona spectrum there cannot
be the least doubt. It is the reversal of a well marked though not
prominent line just below E, put down as *iron* by both Kirchhoff
and Ångström, though not given by Huggins. Whether the position
of the apparently coincident Aurora line is as well settled I am
not yet certain, but hope to be able to ascertain before very long
by a direct comparison between the spectrum of the Aurora and
that of the electric spark from iron electrodes. At present it
seems pretty likely that the spectra of the corona and the aurora
borealis are identical with only such differences in the intensity
of their lines as we might naturally expect, and that very probably
the identity extends to the essential nature of the phenomena
themselves.

Should it turn out that this line in the aurora does actually
coincide with 1474, it will be of interest to inquire whether we
are to admit the existence of *iron vapor* in and above our atmosphere, or whether in the spectrum of iron this line owes its
presence to some foreign substance—probably some occluded
gas as yet unknown, and perhaps standing in relation to the
magnetic powers of that metal.

[The above article comprises the substance of two papers read
at the last meeting of the American Association for the Advancement of Science at Salem.]

9

Notes on Recent
Progress in Solar Physics†

J. Janssen

In this paper we want to bring the readers of the *Annuaire* up to date with the progress that has recently been made possible in the field of solar physics due to photography.

It would be a mistake to attribute solely to the technique of spectral analysis the rapid developments which continually take place in astrophysics, and this will be truer still of future developments. Undoubtedly that excellent method of investigation plays the most important part in such progress, but it is not the only one which can or should be employed. The future of astrophysics rests with all discoveries made in physical, chemical and geological sciences. It must be able to draw equally from all these sources according to its needs, and then by selective assumption it will continue to widen its horizons.

In this context our readers will appreciate the importance of photography to astrophysics, particularly its use in the observation of hitherto unnoticed phenomena.

Before proceeding, however, we will summarize the present state of knowledge about the composition of the Sun.

The best way to indicate the increase in our knowledge about our central star is to remember what conclusions had been reached by the most eminent astronomers only forty years ago.

In his *Astronomie populaire*, vol. 2, p. 181, Arago says, "If anyone asked me, 'Is the Sun inhabited?', I would say that I had no idea. But if anyone asked me if the Sun could be inhabited by a

† *Annuaire du Bureau des Longitudes*, 1879, p. 623.

civilisation like ours, I would not hesitate to say, yes." Such a reply would be almost ridiculous today.

Here Arago adopted Herschel's theory, that the solar globe comprises a dark and relatively cold nucleus, surrounded by an atmospheric layer of highly reflecting clouds, which reflect and repel the heat and light from an outer luminous layer, the photosphere, so called because that is what gives the Sun its property of illumination and makes it a radiant globe.

It was precisely this concept of a dark nucleus, which was cold by virtue of its darkness, which led to the idea of the possibility of life on the Sun. We will see presently what one should think of such ideas.

In 1860, however, the technique of spectral analysis was established, and as a result solar physics was transformed.

Observation of the Sun's light by a spectroscope soon shows that most of our metals are present in vaporized form, either in the bright envelope itself or in the atmosphere above it: the first, decisive step towards the material unity of the solar system.

After the Sun, consider the stars. These distant suns also contain the Earth's metals in various compounds. Thus the unity of the universe is demonstrated.

At the same time, however, it is possible to record a fact of which not enough notice was taken, and which, if properly interpreted, would have permitted anticipation of the discoveries of 1868 concerning the nature of prominences and the existence of the chromosphere: this fact is that most of the stars are surrounded by a vast atmosphere of hydrogen. Using the clear analogy between the composition of our Sun and of those scattered in space, is it not very probable that our central star contains hydrogen as the principal element of its gaseous envelope?

That discovery was made in 1868, during the major eclipse in August, when the French astronomers went to India to observe the eclipse under the most favourable conditions, thanks to the support of Mr. Duruy (the former minister who commands lasting respect in liberal France) and thanks also to the Academy of Sciences and to the Bureau des Longitudes.

The state of knowledge about the Sun has taken a great step forward. The nature of the prominences is understood. They are real objects, and not tricks of the light. They are huge jets of gas, made up mainly of incandescent hydrogen, which rise to heights of 10,000, 20,000 and 30,000 leagues, that is about one quarter of the radius of the Sun.† Soon afterwards it was discovered that these prominences depend on an atmosphere of hydrogen, between 10 or 15 seconds of arc in height, which completely surrounds the Sun. But just as the jets in the prominences rise above the chromosphere, and disperse in the coronal atmosphere, so the chromosphere has eruptions of metallic vapours, mainly of magnesium, which penetrate it from time to time.

The success of spectral analysis does not end there, however. One knows, in fact, that the phenomenon of a total eclipse owes its spendour more to the magnificent aureole of light which surrounds the eclipsed Sun than to the prominences. This aureole or corona, with its surroundings ray, its columns and all its luminous effects, sometimes seems to take up three or four times as much of the sky as the Sun itself.

This phenomenon, however, is as puzzling as it is splendid. Every time a total eclipse has enabled us to study it, it is seen to be accompanied by formations which are so irregular, so bizarre and so fluctuating that their cause cannot be discovered by the ordinary methods of optics. It is spectral analysis, combined with the use of the polariscope, which has enabled us to penetrate—at least to a large extent—the enigma of the corona.

In particular the corona has been examined by a spectroscope in 1869, and at later eclipses.

In 1869, the eclipse occurred in North America. American scientists recorded some very important observations, and mention should be made of their photographs of the corona, which demonstrated the considerable actinochemical power of the phenomenon, and of the confirmation of the green ray (Kirchhoff's maps,

† 2 leagues = 5 miles.

1474), which appears to be characteristic of the coronal spectrum.

Another eclipse took place the following year in the Mediterranean basin. This time scientists from most of the civilized countries took part in the observations. Many posts were established for the study of the phenomena, in Sicily, Africa and Spain.

At that time, France was an invaded nation, and Paris was under siege. Many of my friends on the other side of the Channel, who wanted me to take part in the observations, very kindly asked Mr. Bismark to allow me safe passage from Paris, and their request was granted, but I had already made preparations which did not depend upon the generosity of our enemies. With the approval of the Government, and under the auspices of the Academy of Sciences, I had prepared my flight from Paris by air. A balloon, similar to those which the Government used for despatches (the *Volta*), was put at my disposal. I took with me a telescope of the latest design, which would provide a spectrum from the aureole some fifteen to sixteen times brighter than that of an ordinary astronomical instrument, and which thus promised to remove the principal difficulties which had been encountered in the analysis of the mysterious phenomenon. That telescope, which had a mirror of 0·04m diameter, was more an observational than a navigational instrument, and the route I was going to follow did not permit easy transport. But I reflected that these difficulties were not insurmountable—the body of the instrument could be replaced, for a brief observation, by a substitute made in Algeria, where I was to go to make observations. Thus I took only the mirror in its mounting, together with all the attachments. These parts were packed and put into boxes full of screwed up paper, which served to cushion them so that even the most extreme shocks could not damage them. The boxes, encircled with iron bands and reinforced on the outside, were placed around the car of the balloon.

I set out on the 2nd of December at 6 a.m., the day of the Battle of Champigny. A sailor travelled with me to help with the navigation, but I piloted myself. We crossed the enemy lines at a height of 800m, but the Sun soon had its effect on the gas in the balloon and we gradually rose to 2000m. The compass showed us the way to Brittany. By 11 a.m., we were at the mouth of the Loire, facing the ocean. A rapid descent enabled us to land in time. We had covered the distance from Paris to Nantes in five hours. A special train took me to Tours, where I saw several members of the Government and also Mr. Thiers on his return from his patriotic mission in Europe. From Tours I went to Marseille and then to Oran, where I was to make observations. I had chosen a station on the outskirts of town, at the Combes tower. An English delegation, including Mr. Huggins, Mr. Tyndall and Admiral Ommaney, had also come to Oran to study the eclipse.

Several days before the eclipse the new body of the telescope was installed, and everything was ready. But fortune did not smile on our dear and unhappy country; it had been raining particularly heavily in Oran for some time. I had sent observers to the provinces of Algiers and Constantine, to increase our chances of success, but in vain; the sky was overcast almost all over Algeria on the day of the eclipse, and it was impossible to make any fresh observations.

In Sicily and in Spain some observations were made whilst the sky was clear. The results were similar to those obtained in the previous year. It is fitting, however, to make special mention of the fine observation by Professor Young, who established that the spectrum was reversed at the base of the chromosphere.

Yet the nature of the phenomenon was still undetermined from these observations. Almost all the observers had found that the coronal spectrum was continuous, which indicated that the corona was made up of solid or liquid matter, and that theory was widely supported. From another point of view, the presence of a luminous ray (1474) and the irrefutable existence of polarization showed, on the contrary, a phenomenon of a gaseous nature.

Such was the state of affairs in 1871, just before a new eclipse which would take place over Asia and Australia. The event aroused lively competition in Europe; France, England, Italy, Holland, etc., took an active part in the observations.

I had the honour of being nominated by the French Government and the Bureau des Longitudes.

Having given a lot of thought to the 1869 observation, it seemed to me that the main difficulties encountered by the observers were related to the poor luminosity obtained from coronal spectra, a fault connected with the inadequate luminous power of the instruments in use, with the result that the detection of faint lines on a bright background, particularly the darker lines, becomes very difficult.

I returned, therefore, to the special telescope I had had made for the 1870 eclipse, in which a very sharp definition had been sacrificed to the all important attribute of luminous power.

The mirror of this telescope has a diameter of 0·40m and a principal focal length of only about 1·60m. In this instrument, the image of the corona becomes about sixteen times more luminous than in an astronomical glass of standard focal length.

This telescope also has a finder, described in my report to the Academy, which enables the observer to view the phenomenon with one eye while the other is applied to the spectroscope, a facility which does away with the need for an assistant and allows one to watch both the event and its spectral analysis.

In addition I had added a polariscope, the readings of which are extremely important for the corona theory.

Thus the preparations for this eclipse were made with the aim of obtaining from the corona a much more luminous spectrum and to combine the results with those from polarization and from the general appearance of the phenomena.

I observed this eclipse at Shoolor, in the Neelgherry mountains (Hindustan), and I was fortunate in experiencing a day of perfect clarity, of a kind I have never seen before or since.

At this point it would be helpful for the reader to study the passage of my report which summarizes and discusses the observation. That report had little publicity, and the last studies made of the corona during last July's total eclipse renders it of current interest.

The Observations

Totality approached. The sky was wonderfully clear. I had set myself a programme, for the total eclipse lasted only 2 minutes.† It was only possible to make a few short observations, but I selected those that could decisively settle any remaining doubts about the nature of the corona.

My main task was to determine as precisely as possible the true nature of the coronal spectrum and if, as I imagined, it displayed the characteristics of a gas spectrum, to determine which gases and what they had in common with those in the prominences. I would finish by examining the results of the spectral analysis to see if they agreed with those of the polarization. First of all, however, I had to view the corona for about 15 seconds in a field glass in order to form an exact impression of the event and to settle on the points which the spectral study was to cover.

The Sun became completely eclipsed; it was reduced to a thin luminous strip which soon changed into separated grains. I let fall the dark glass of the field glass, and the corona appeared in all its splendour.

Several prominences, of a rose coral shade, shone around the Moon, against the background of a faintly luminous matt-white velvety aureole.

The contour of this corona was irregular, but clearly defined. The general shape was that of a curvilinear square centred on the Sun, its corners reaching away to a distance of about twice the radius of the Sun, the edges extending only about half the Sun's radius.

Neither diagonal was situated along the solar equator. The corona had a most curious structure, which will prove useful in settling several theoretical questions. It was possible to distinguish several luminous trails leaving the Moon's limb to join up with the upper parts of the corona. The appearance is that of a pointed arch or of the petals of a dahlia. This formation was repeated all round the Moon, and altogether the corona was like an enormous

† Calculations give 2 minutes 6 seconds for the length of the totality at Shoolor.

luminous flower with the centre taken up by the Moon's black disc.

I threw off my momentary state of ecstasy produced by this incomparable phenomenon in order to pursue my work. I looked to see if there were any essential differences between the appearance of the corona at the point of contact and at the opposite point. I could find no difference. I then followed the eclipse for a few moments, in order to see if the movement of the Moon would cause any important changes in the initial form of the corona, but nothing like that happened. Thus I was completely convinced that I was looking at the image of a real object which was situated beyond our Moon, the various parts of the object being revealed by the progressive motion of the latter.

Having completed that investigation, I returned to the luminous elements of the phenomenon. Since my vision was still quite acute, I began an inspection of the spectrum of the highest and least luminous sections of the corona. I aimed the slit of the spectroscope at a point in the corona about two-thirds of the Moon's radius from the edge of the Moon. The spectrum displayed much greater brightness than I had expected at that point, obviously thanks to the high power of the instrument and to the various components that were being used. The spectrum was not continuous. I was quickly able to distinguish hydrogen lines, and the (1474) green line.† This is a point of primary importance. I moved the slit, keeping within the upper region of the corona, and the spectrum still showed the same composition.

From one of these positions, I then slowly moved it down towards the chromosphere, keeping a careful watch for any changes that might take place. As I approached the Moon the spectra appeared to get much more active, and richer, but the general composition did not change. At an average height of the corona of 3 to 6 minutes of arc, the dark D line was visible, as were a few dark lines in the green region, but these were at the limit of visibility. This observation showed that reflected solar

---

† My spectroscope had a very exact scale, but readers will see later how I used the prominence lines themselves as a scale.

light was present in the corona, but it seemed that this light was submerged in a strange and abundant luminous emission.

I then made a start on the most important part of the observation, which was to provide me with the spectral relationship between the corona and the prominences. The slit was positioned so as to take in a part of the Moon, one prominence and the full height of the corona.

The prominence had a very rich spectrum of great intensity; there was no time for a detailed study of it. The main point here was to establish that the principal lines of the prominence extended to the full height of the corona, proving without question the existence of hydrogen there.

The (1474) green line which was so strong in the spectrum of the corona appeared to be broken up in the spectrum of the prominence, a very remarkable result. I spent a few moments establishing the exact relationship between the lines of the corona and the main hydrogen lines of the prominence.

I only had a few seconds left to use the polariscope.† The corona displayed the characteristics of radial polarization and, more important, the maximum effect was not observed at the base of the Moon's limb but at a few minutes of arc from the edge.‡

I had hardly finished this rapid observation when the Sun reappeared.

Discussion

In the study of a phenomenon as complex as the corona, it is necessary to combine various methods of investigation. Hence I considered it essential to examine the corona from three points of view—from its appearance, from its luminous analysis, and from its polariscopic properties. We will discuss these various observations.

Firstly, let us see what we can learn from the examination of the corona during the first moments of totality.

† To study the polarization I used an excellent glass with a biquartz, which was made by Mr. Prazmowski. If the glass is placed on the telescope and aligned it can be used at a moment's notice.

‡ Mr. Prazmowski indicated this fact in his polariscope observations of the eclipse which took place on July the 17th, 1860.

We saw that the general structure of the corona remained constant throughout the eclipse.

It would be invalid to invoke here an effect such as diffraction, caused by rays touching the edge of the lunar disc. In fact, let us return to the geometry of a total eclipse. At the precise moment of totality the discs of the Moon and the Sun have a common tangent, and then the Moon's disc goes beyond that position until the opposite edges are in line. Diffraction would take place therefore in the most different physical conditions at various points on the Moon's limb, and the resulting aureole would show, by its lack of symmetry, the differences in those conditions.

Moreover, this kind of aureole undergoes continuous change during the different phases of totality. At first, it is asymmetrical, and then tends to assume a shape similar to the Moon's when the latter hides the Sun symmetrically. Finally, from that position, the aureole goes through all these phases in reverse until the Sun reappears.

Nothing like that happened at Shoolor, where the general form of the corona remained the same throughout totality.†

As for the hypothesis of an aureole produced by the moon's atmosphere, this need not be considered in detail. If a gaseous layer does exist over the surface of our satellite, we now know that it can only be of negligible extent, incapable of causing the impressive spectacle of the corona.

Our own atmosphere can no longer be regarded as the cause of the phenomenon, although it evidently has a part to play in the formation of the particular shapes of the corona which are seen from various observation posts, as a result of prevailing weather conditions. Thus it modifies, but is not the cause of, the phenomenon.

Let us go on to the spectroscopic observations.

The corona displays the hydrogen lines to the extent of its

† It is, furthermore, clear that this constancy of appearance refers only to those points of the general structure which are sufficiently far from the Sun not to be influenced by the variations in brightness due to the displacement of the Moon relative to the deep, very luminous regions of the chromosphere.

visibility, in some places up to a height of 12 or 15 minutes of arc.

This is a precise observation; the accuracy of the spectroscopic scales, our familiarity with making such observations, and the care taken over them during the last sighting, in comparing the lines of the corona with those of a prominence of which they were extensions, all these serve to endorse this fact.

If the corona exhibits hydrogen lines, however, we must then consider a vital question: Is the light emitted or reflected? The composition of the coronal spectrum provides the answer.

If the light from the corona is reflected light it can only come from the Sun: it comes from the photosphere and the chromosphere and its spectrum ought to be that of the Sun, namely a luminous background with dark lines. The coronal spectrum does not have this kind of composition, but displays hydrogen lines which show up clearly against the background; apart from the (1474) green line this is the most significant characteristic of the phenomenon. Thus we must conclude that the coronal medium must emit its own light for the most part at least, and must comprise incandescent hydrogen.

This first point is clearly established. But does this mean that all the light from the corona is emitted light? Obviously not, and on this point we can find out more from a careful analysis of the spectrum and polarization of the light.

Apart from its bright lines, the spectrum of the corona actually showed several dark lines of the solar spectrum, the D line and some in the green. This indicates the presence of reflected solar light. One might wonder why the D line is the only one of the Fraunhofer lines which has any prominence here. The reason is that, since the coronal spectrum is not highly luminous, it is most distinct in the central region, and in that region the lines C, F, etc., give rise to bright lines. In these conditions, the D line was the only important one which remained—it was, therefore, that one to which I directed most of my attention. As for the finer lines, they were much more difficult to make out, which is explained by the fact that I had to increase the aperture of the spectroscope.

The detection of Fraunhofer lines in the coronal spectrum is tricky, and has not been done by other observers. I was able to do so thanks to the perfectly clear sky at Shoolor and the power of my instrument. I anticipate that other astronomers who experience similarly favourable conditions will confirm that observation.

The presence of reflected sunlight in the coronal spectrum is of great significance: it shows the two-fold origin of coronal light; it explains the seemingly incompatible polarization observations;† above all, given that the Sun's light also forms the background of the coronal spectrum, it explains how one can believe in this continuous spectrum. Up to now, this had been the main objection to the idea that the corona was composed entirely of gas. The polarization phenomena displayed by the corona are predominantly those of radial polarization, which shows that the reflection takes place mainly within the corona, and that produced in our own atmosphere is only secondary. Thus the polarization is in accordance with my remarks on the Fraunhofer lines; in order to show that the agreement is complete, however, the polarization analysis must demonstrate, just like the spectral analysis, that only part of the light from the corona is reflected light. This is exactly what happens. We have seen that, near the Moon's limb, where the coronal light is most vivid, polarization is less pronounced than elsewhere. The reason for this is that the emission is so strong in the lower regions that to some extent it masks the reflected light, which only shows up properly from the layers in which it has relatively more strength.

Thus the correct interpretation of the polarization and spectral analysis shows general agreement concerning the dual origin of the coronal light, and all the observations combine to establish the existence of a circumsolar medium.

This medium is distinguished by its temperature and by the chromospheric density, and the shapes of the prominences and

† If one looks at the history of eclipses, one can see that observers have often obtained contradicting results, which has somehow discredited this kind of observation; however, one can get rid of most of the difficulties by discussing these observations in the light of the two-fold origin of the coronal light and of the effects of our atmosphere.

the chromosphere serve to define its limit precisely. It can be given a name, therefore: I suggest *coronal envelope* or *atmosphere*, in order to show that the luminous phenomena of the *corona* originate in it.

The coronal atmosphere must have a particularly low density. In fact, the spectrum of the upper chromosphere shows it to be highly rarefied hydrogen, and since the spectral indications are that in the coronal region it must be much less dense, then this indicates the high degree of rarefaction attained there. This conclusion is also corroborated by astronomical observations. Comets have been seen to pass within a few minutes of arc from the surface of the Sun; they must have passed through the coronal atmosphere and yet, in spite of their low mass they have not fallen on to the Sun.

Regarding the composition of the coronal atmosphere, I should like to describe several other ideas which have not been derived with any particular rigour from my observations, but which seem to me to be most feasible and, moreover, can be decided by future observations.

I have described how the corona which I observed from Shoolor was nearly square in shape, and that parts of it looked much like the large petals of a dahlia. It is a fact that the shape of the corona is different at each eclipse: it often looks most bizarre. I should say immediately that this medium, which is now accepted without question and which I think should be called the *coronal atmosphere*, probably does not comprise the complete aureole that we see during total eclipses.

According to the theories of Mr. Faye it is quite possible that parts of asteroid rings or trails of cosmic matter may become visible and will thus complicate the shape of the corona. We will learn more about this from future eclipses. But in considering only the coronal medium, it definitely does assume some strange shapes, which do not at all suggest that it is formed from an atmosphere in equilibrium. I must remark that such shapes are brought about by trails of denser and more luminous matter from the lower layers which furrow into the turbulent regions.

ı such phenomena an important part must be played by the jets from the prominences which take the hydrogen up to great heights. It will also have to be established whether the Sun, which has such obvious effects on comets, has any particular influence on the coronal medium, the density of which compared closely with that of cometary media.

It is highly likely, therefore, that the coronal atmosphere, just like the chromosphere, is very turbulent and changes its form quite quickly, which would explain why it assumes a different appearance every time it is observed.

To sum up, I was able to establish by certain concurrent observations at Shoolor that the solar corona exhibits the optical properties of incandescent hydrogen, and that this rarefied medium extends to widely varying distances from the Sun, from about half the Sun's radius up to twice its radius in some places, where it would have a height of 80,000 to 160,000 leagues (of 4km each); but these figures result from one observation only, and are therefore not final. What is certain is that the height of the corona is subject to continuous change.

This seems to mark a considerable step forward in general investigations of the corona. Our colleagues overseas may not have obtained such positive results† as those of the French group, but I think the reasons must be that the sky was so exceptionally clear over the site of our carefully chosen observation post, and that the phenomenon we wanted to examine was shown up to great advantage with the help of the ancillary optical instruments.

From this it is apparent that this observation established quite clearly that the coronal phenomenon took place in a gaseous medium. I suggested that it should be called the *coronal atmosphere*, a name which has now been generally adopted.

As for the continuous spectrum, seen in 1869, 1870 and again in 1878 (and also by me in 1875), it could be due to solid cosmic

† At Poodoohotali, Mr. Rispoghi made excellent observations of a purely spectroscopic nature which confirm my observations; however, he found that the height of the corona was much lower than my values, but I attribute that to the weaker power of his instrument.

matter circulating more or less freely round the Sun, in accordance with Mr. Faye's ideas about the subject.

Here, I should like to draw the reader's attention to a theory which I put forward at the Congress of the British Association in Glasgow, and at this year's congress in Dublin. It seems reasonable that the shape of the corona should be related to the number and the strength of the prominences which extend into it. At times when a minimum number of spots is visible, which is when there is also a minimum number of prominences to be seen, the shape of the corona must tend to be that of an atmosphere in equilibrium; but at a period of maximum activity of the prominences, when they penetrate the corona in all directions, it must have a much more irregular and disturbed shape.

If we had accurate maps of the corona during the various total eclipses which have been observed, then we could see if this assumption had any foundation. Unfortunately the drawings and descriptions which we do have leave much to be desired. We ought to point out, however, that in the case of the eclipse of 1842, which coincided with a minimum spot period, and which was described so well by Arago's party, we can tell that the corona had a regular form, quite concentric with the sun and even made up of two quite distinct rings, which was definitely typical of a state of relative calmness and equilibrium.

In this survey, however, it should not be forgotten that the cosmic matter, the swarms of meteorites travelling in the vicinity of the Sun, can complicate the picture, and at least partially hide the true aspect of the phenomenon. In future, scientists will be able to distinguish between this solid matter and the true material of the corona.

To sum up, the value of spectral analysis in investigations of the Sun's gaseous envelope (disregarding for the time being the vexed question of zodiacal light) can be said to be that this technique has enabled us to distinguish three quite distinct envelopes.

Immediately above the photosphere—the envelope whose luminous power places it above the others, and which gives the Sun its radiant power—there is a very thin layer of hardly a few seconds of arc comprising incandescent metallic vapours lighter than those found in the photosphere; above that is the chromosphere, of true height 8–12 seconds of arc, a very hot layer in which hydrogen predominates, but which has frequent injections of metallic vapours of magnesium; finally there is the coronal atmosphere, a very high, greatly rarefied region, much cooler and very turbulent, an atmosphere which is rarely in equilibrium and where the effects of cometary matter approaching the Sun make themselves felt. All this, plus the less frequent appearance of rings and of meteorites, combines to produce the appearance (and strange shapes) of the solar envelope which for so long astronomers have found puzzling.

So much for recent discoveries.

Let us now go back to the photosphere.

We know that this envelope, which serves to make the Sun a radiant star, has had its nature under scrutiny ever since the invention of the lens, and that this investigation (a principal activity of astronomers for more than 250 years) has revealed the rotation of the Sun, the inclination of its axis relative to the ecliptic, the structure of sunspots and their various conditions of movement, the changing speeds of photospheric zones according to their different latitudes, and so on. All these wonderful observations were made with the help of astronomical glasses and telescopes.

Spectral analysis, which has laid before us so much information about the outer envelopes, has not contributed so much about the photosphere. This is because there are a variety of different techniques, each of which has its own particular difficulties to be overcome.

It seems that the wonderful French invention of photography will help us take the most decisive steps in learning about the photosphere, and we are about to embark upon a new phase of our studies of the Sun.

Before telling our readers of this new application it seems necessary to us to give a theoretical discussion of the photographic method as compared to simple visual or telescopic methods; this discussion will make it easier to understand our final conclusions.

Let us say first of all that bearing in mind the wide range of applications of photography, its value to all branches of science and industry, and also more important future applications, it seems imperative that astronomy should immediately take advantage of this excellent method for the recording and detection of phenomena. We are convinced of the potential value of photography to astronomy, and that has persuaded us to continue these studies in solar photography.

In this connection, and in order to show the advantages of photography, we will outline the similarities between the visual image and the photographic image.

First of all we will consider the relative ranges of visual and photographic spectra.

The visual spectrum extends from red to violet.

The photographic spectrum extends at least as far as the visual spectrum towards the red, thanks particularly to the work of Messrs. Becquerel, Vogel, Waterhouse, Abney, etc. As for the violet end, it is well known that photographic plates are sensitive to rays situated beyond the violet, not visible to the eye under normal conditions, which are consequently known as *ultraviolet* rays.

This part of the spectrum has been studied in France by Messrs. Becquerel, Mascart and Cornu. In a recent experiment, Cornu succeeded in photographing solar rays having a wavelength of only 0.000295mm whereas the wavelength of violet light commonly regarded as being at the limit of visibility is about 0.000390mm; the range of the photographic spectrum is thus about one-third greater than that of the visual spectrum.

Thus photography can reveal some phenomena, in which ultraviolet rays take part, which would escape human vision.

But the advantages do not end there.

A visual image has an intensity which is limited by the duration of the impression of light on the retina. The intensity of the image formed at the back of the eye does not gain strength after about one-tenth of a second, because the sensations of light fade away after approximately that duration and new ones replace them. The gain does not exceed the loss. The intensity of the image seen by the eye is regulated by the duration of the impression of the light on the retina. If it were natural for the impression of light on the retina to last for one-fifth of a second, the luminous effect would be doubled and the world would appear to be twice as bright. In other words, it would be twice as difficult to endure daylight, but at night it would be possible to see twice as well into the starry depths. A whole collection of stars which escape the naked eye would have been revealed to us before the invention of the telescope. If the image duration had been 1 second the world would be ten times as bright, daylight unbearable, and the night would shine with such an enormous number of stars that the sky would seem like one huge Milky Way.

An extension of the duration of light persistence on the retina would have that kind of effect. The effective length of the light impression on the eye is fixed, however, and fixed within fairly narrow limits. In photography exactly the opposite is true, and the light has a cumulative effect, with nothing to limit its action. Using plates of the kind available today, with dry collodion, one can record, over a comparatively unlimited time, weak luminous effects. Impressions on the eye are gone almost as soon as they are formed, but those which fall on a sensitive plate are not lost and will combine with those which it receives subsequently.

Not only does the photographic eye have a much more extensive recording capacity than the human eye, retaining the impressions made on it, but also it has the property of compounding light received over a period of time to brighten objects which otherwise would appear lacking in brightness.

Let us pursue this similarity a little further.

We know that the yellow region is the brightest part of the solar spectrum. Next in brightness comes the orange, then scarlet, green and blue; and finally the extreme points of violet and red, one of which borders on purely thermal radiation, the other the so-called chemical radiation, which has been revealed to us by fluorescence and photography.

This same sequence is displayed in another way when spectral light fades at the end of the day. If one considers the spectrum of the clouds and watches the colours pale one by one, the violet disappears first of all, then the blue; the red contracts and becomes darker, then the green vanishes and finally, before disappearing altogether the spectrum is reduced to a whitish band which takes the place of the yellow.

In conclusion, the human eye is more sensitive to yellow light than to any other, and it is necessary to ensure that the luminous elements of the solar spectrum that are under comparison (within the limits of visual sensitivity) reach the retina under conditions of equal radiating power: although the method of ensuring this has not been properly explained, it is not essential to our purpose, which is to determine the upper limit of light visible to the human eye in the solar spectrum. What we really want to show our readers is that a maximum exists similarly in the photographic solar spectrum, although it is situated in quite a different region and has some most unusual properties.

We have seen that, for the camera, the solar spectrum can extend from the extreme red to well beyond the violet which is visible to the naked eye; at the same time it would be wrong to imagine that the spectrum is of constant intensity over its entire length. A cursory examination of the photographic image of the spectrum, or of the opaqueness of the metallic deposit which forms the image, is enough to show that the intensity varies in different parts of the spectrum.

The image is most intense in the violet region, and the intensity decreases as one goes towards the red of the ultraviolet.

In 1874, while I was considering these spectra in connection with preparations for an expedition to Japan to observe the passage of Venus, it occurred to me (as it probably has to many others) to study the behaviour of the photographic image of the solar spectrum by gradually reducing the time of action of light to a point where the image ceases to be formed.

I found, as anticipated, that the weak parts of the image were

E.S.P.—6

first to vanish—in other words the photographic image contracts as the time of exposure is reduced.

But a result which was least expected was that the spectrum contracts to a narrowly defined, clear-cut band when the exposure time is sufficiently reduced. The narrow band which represents the fading spectrum is near to the G band.

Below is an illustration showing the various lengths of the photographic solar spectrum when the time of action of light is successively reduced.

Figure 1 shows the solar spectrum in which the modern techniques have produced the yellow and the red.

In Figure 2, the spectrum is still rich in the ultraviolet but does not extend beyond the b line in the other direction.

In Figure 3, the spectrum stretches from F to P.

In Figure 4, the spectrum is reduced to the region between G and L.

Figure 5 shows the spectrum concentrated into the band mentioned above, which indicates the point with the maximum photographic effect.

Thus each spectrum has a maximum, but in the photographic spectrum the maximum is much more limited and well-defined than in the visual spectrum. This is why the photographic spectrum reduces to a band as it fades.

We have studied the various salts of silver, zinc, lithium, etc., from this point of view; the maximum was situated at more or less the same place for each.

This property of the photographic spectrum is important in connection with the achromatism of the lenses to be used in solar photography. We know that these images need quite short exposure times, and in such cases the rays under consideration are very close together, so that it is possible to determine the achromatism with a precision that is not feasible for eye-glasses.

Let us consider further another of the advantages of photography, that of the extent of the field of the photographic image. We know that the greater the power of a telescope, the more restricted is the field of effective vision of the observer. If an

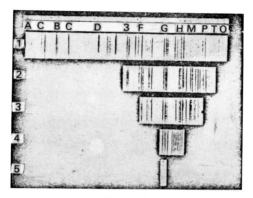

Figs. 1–5.

astronomical telescope, of magnification × 30, shows the Sun occupying the whole field of the instrument, then under twice that magnification the same telescope shows only half the height of the disc and one quarter of its area, four-fold magnification giving one sixteenth the area, and so on. When the instrument in use is high-powered, combining a very great focal length and a powerful eyepiece to produce a magnification of 1500 to 2000 or more, then that part of the Sun which appears in the viewfinder could be less than 1/3000 of the area of the solar disc. The detailed structure of such a small area can be examined closely under such conditions, but adjacent phenomena disappear completely and so comparisons of simultaneous phenomena are just not possible. In photography, things are quite different: greater magnification does not necessarily reduce the field which can be examined. Pictures of the whole of the solar disc can be obtained, ranging from a diameter of a few millimetres to the largest size required. The major difficulty concerns the handling of the larger sizes of photographic plate.

Solar images of diameters 10, 20, 30 and 50 cm have been obtained at the observatory at Meudon, and it will soon be possible to produce an image of 70 cm diameter. It is most interesting to enlarge the images in this way; in producing a highly detailed image of the phenomena which are taking place simultaneously on the surface of the Sun it is possible to observe relationships which would otherwise escape if one used successive observations which is what a powerful telescope must do. It will be seen that it is thanks to this aspect of the photographic method that the photospheric network on the Sun's surface was discovered, a system which had escaped astronomers' observations for more than two and a half centuries.

There are yet more comparisons to make between photography and human observation, particularly in connection with the accuracy of the respective images and the extent to which they present the true intensity of those parts of the object under observation.

We also have to consider the techniques for recording the

images, in which context the names of Arago, Fizeau and Foucault are of note, and that subject is worthy of further study.

These are some of the general comparative features of human vision and photography, and they indicate the many advantages of this technique which are likely to be beneficial to science.

Let us now consider solar images.

We know that so far solar photography has been unable to reproduce some of the details shown by powerful instruments. The most remarkable photographs of the Sun produced to date (among them those of Mr. Warren de la Rue, of Mr. Rutherfurd, etc.†) showed quite clearly the spots and the faculae; but for the surface itself they show none of the granulations which can be seen through optical instruments, but generally give only a mottled effect.

It should be remembered that it was not anticipated that photographic processes would be capable of reproducing such fine details, of the kind which are normally observed under the most favourable weather conditions.

Eventually I concluded that inferior techniques were responsible for that shortcoming, and that it was not a basic defect of photography itself.

Indeed, after some deliberation, I have decided that for some effects and relationships of light phenomena to which the eye is insensitive the techniques of photography are much better for purposes of detection and observation.

The human eye has a wonderful ability to function under the most varied conditions of illumination, but our vision is not capable of measuring comparative intensities of illumination, particularly at very high intensities.

And this is the case with the solar image. In spite of the introduction of coloured glasses, helioscopes, etc., the eye still has to pick out details in the dazzlingly bright regions, and to function under conditions quite alien to it. It is then impossible to make

† In 1849, Messrs Fizeau and Foucault obtained a photographic plate of the Sun which was apparently the first image of a celestial body produced by photography.

true comparisons of the intensities of light from various parts of the image, and appearances no longer coincide with fact. The widely differing descriptions of the shapes and sizes of the granulations and constituent parts of the Sun's surface are probably due to that reason.

When the photographic image is obtained under well-regulated conditions of light action it does not suffer from these faults, and provides a fairly good approximation to the light intensity of the various parts of the object which created it.

If this valuable result is to be produced, it is essential that throughout the action of the light, the sensitive layer on the plate remains more or less unchanged—in other words, that portion of the photographic substance which is likely to be affected by the light at any time during the exposure should be in weak solution on the plate.

I shall have to come back to this highly important point.

Thus, by regulating strictly the time of light action, so that there is no *over-exposure* of the brightest parts of the solar disc, it is possible to obtain an image which will show us not only the true details of its contours, but also indicate very approximately the true relationships between their luminous intensities.

Photography has yet another very valuable advantage over human sight, especially when exposure times are short. I found (as mentioned above) that when the duration of the light's action is made much shorter than normal, then the photographic spectrum is reduced to a very narrow band situated near G.

This unusual characteristic indicates that it is possible to obtain a very acceptable photographic image of the sun by using a simple lens of considerable focal length. Above all, it shows that it is much easier to achieve chemical achromatism than optical achromatism, and that, in particular, a solar image thus obtained can have a much greater precision than an optical image.

These are the advantages, which I call technical advantages, that photography has over the optical method. The inferiority of some solar photographic images obtained up to now is mainly due to the unfavourable conditions under which they were taken.

First of all it is necessary to study the action of light during very long exposures.

In fact, when the action of light is too long in relation to its intensity, the photographic image increases rapidly in size and loses all definition. This phenomenon, which could be called *photographic irradiation* (without prejudging its origin), is most striking in the photographs which have been taken at total eclipses since 1860. On such photographs, the prominences are seen to encroach upon the lunar disc by as much as 10 or 15 seconds of arc or more. Obviously, in the study of solar granulations of diameter of 2 to 3 seconds of arc on average, it is not possible to detect them on images where photographic irradiation produces an effect of larger dimensions than those of the granulations.

In order to overcome this major obstacle, I made a very careful study of the time during which the light acts in accordance with the previously stated principles.

I combined a decrease in time of light action with an increase in the size of the images. The sizes of the images were successively 12, 15, 20, 30 and 50 cm.

The time of light action, which in this case is an important factor for success (since images of parts of a solar disc of effective diameter 1m have been produced which have shown no granulation) has been reduced to 1/3000 of a second in summer.† A special high-precision mechanism is necessary to produce accurately such a short exposure and to create an equality of luminous effect to within 1/10000 of a second for the different parts of the image.

When the exposure time is so short, the image is even harder to obtain than under normal circumstances; it has to be developed slowly, and then strengthened with pyrogallic acid and silver nitrate.

Needless to say, the photographic processes have to be performed with the greatest care, when it is intended to produce images which will show such fine details. In particular the gun-

† The figure relates to the action of natural solar light which had not passed through any refractive medium.

cotton must be prepared at a high temperature if a sufficiently fine layer is to be formed. Given these conditions, it is possible to obtain solar images which, in comparison with previous ones, represent a whole new world of experience, and display phenomena which we will now consider briefly.

First of all, however, I must mention that the camera lens that I used in the research was specially made for our expedition to Japan by Mr. Praxmowski, who based his lens calculations on the spectral data which I supplied, concerning the maximum effect which I have already discussed.

In the photographic work I was very ably assisted by Mr. Arents, a photographic expert at the Meudon observatory.

Photographic Processes

For those of our readers who would like to take solar photographs which will show surface granulations, we give below a few details of the process.

It is not possible to over-emphasize the need for the absolute cleanliness of the plates; cleanliness is much more essential here than it is for the most accomplished artistic photographs.

The plates must be passed through potash, nitric acid and polishing powder, and polished with a wad of cotton wool moistened with a few drops of ether sharpened with acetic acid.

An extreme fineness in the texture of the collodion is most vital. The method of preparation is as follows:

Preparation of gun-cotton

Sulphuric acid	500g
Potassium nitrate	500g
Cotton	15g

First pour the sulphuric acid into a porcelain dish and then gradually add small amounts of potassium nitrate, taking care to stir with a glass rod until all the grains have disappeared and a homogeneous mixture results. The temperature at this stage can be about 60°. Heat gently to 80°, then place the dish in a heated bath of sand to maintain that temperature. The 15g of cotton, which has been shredded, should then be added very gradually, care being taken to ensure that it is well strung out and pressed into the dish so that the liquid is thoroughly absorbed by the cotton. It is important that the cotton is added in small, well-separated strands, because if it is put in all at once, it would retain air in its fibres, would give off dark yellow nitrous fumes, and the resulting cotton would be defective. Such gun-cotton would be highly soluble in alcohol, and would produce very intense images, but it would be very difficult to fix it. The cotton should be immersed in the liquid for about 6 minutes; from experience, and according to the type of cotton used, it is possible to tell if more or less time is required. In any case, it should never be more than 10 minutes, unless the liquid has cooled below 60°.

When the cotton has been left long enough to soak up all the acid, a considerable quantity of water is added, care being taken to stir briskly in order to dissolve the salt which sticks to the cotton fibres. When done under favourable conditions, this washing process leaves the cotton soft and pliable, an indication that there is no potassium sulphate remaining in it. The washing should be continued for several hours, until the water gives a neutral reaction when tested with litmus paper. The cotton is then wrung out, and the tufts separated and allowed to dry in the open air. Guard against grains of dust settling on the cotton—such dust, if it stuck to the cotton, would show up after the development process as spots rather like comets (round dots with tails); grains of dust always remain in suspension in collodion. Gun-cotton which has been prepared properly should weigh about a quarter more than it did to start with.

Collodion

Alcohol at 40°	400g	Cadmium iodide	5
Ether at 62°	600	Potassium iodide	2
Gun-cotton	15	Ammonium bromide	1
Ammonium iodide	4	Cadmium bromide	1†

The collodion must be spread very evenly over the plates. The silver baths must be scrupulously clean and should always be quite new.

† Mr. Arents.

After exposure, which varies according to the season from 1/500th second to 1/3000, 1/4000 and even 1/6000, and which is regulated by a special mechanism that we are going to describe, the plate is developed slowly by the iron process, and then very carefully washed and strengthened with pyrogallic acid and silver.

Fig. 6. Apparatus for regulating exposure time.

A. Base for the mobile apparatus *BCB* which slides on the four rollers *r*.

BB. Moving part carrying the shutter.

C. Small plate operated by the screw *v*, which also controls the aperture of the shutter.

f. Thread which keeps tight the springs on the mobile part.

P. Button for securing the thread.

The apparatus in Fig. 6 is placed in the body of the telescope, at the focus of the lens, so that the real image of the Sun given by this lens is formed in the circular opening in the plate of the apparatus, near its centre.

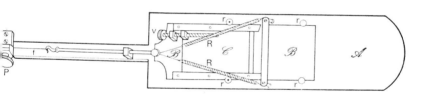

Fig. 6

If the apparatus was reduced simply to its flat base, then the light from the real image would pass through the aperture, fall on the finder, and form an enlarged image of the Sun on the photographic plate. The base carries a moving part *BCB* which slides on the rollers *r*, and a composite part of this mobile section is a shutter which can be opened by varying amounts by means of a small movable strip *1* controlled by a screw *v*. At one end of *BCB* there is a rod with a hook, *f*, and attached to the hook is a thread which can be secured at *P*. The moving part *BCB* is connected to the bridge *p*, on the base, by springs, which are in tension when the moving part is in the position shown in the figure. If the string is cut the springs pull the moving part to the right, over the rollers. In that instant the shutter passes over the circular hole in the base and allows light from the various regions of the real image to pass through and print their images on the sensitive plate. Thus the various parts of the image are photographed in succession. The regular movement of the shutter is extremely important, and I actually achieve this result by using a device in which the springs only serve to give the mobile plate its initial impetus, and their action ceases as soon as the image begins to form. Thus the mobile plate moves by virtue of the speed it acquires, and its motion is uniform throughout the very short time of exposure.

The exposure time for each part of the image is easily measured with the help of a tuning fork. A small piece of blackened paper is attached to the mobile plate, and a stylus fixed to a vibrating tuning fork touches the paper during its motion. The result is a wave on the blackened paper, from which the velocity of the plate can easily be determined. That velocity, together with predetermined shutter opening, permits the calculation of the exposure time. We still relate the exposure time to the direct solar light, in other words sunlight which has not been subject to intensification or weakening.

We will now consider briefly what these photographs can convey about the composition of the photospheric layer.

As we have already seen, the photographs show that the solar surface is covered with a general granulation. The shapes, sizes and distribution of the granulation are not in accord with the results of the optical observations of the photosphere. The theory that the photosphere is made up of elements whose fixed shapes are like willow leaves, grains of rice, etc., is in no way confirmed by the photographic images.

These shapes, which do occur at various places, are exceptions and cannot be regarded as demonstrating any kind of general law for the composition of the photospheric region. Photographic images suggest much simpler and more rational theories about the make-up of the photosphere.

Shapes of Granular Elements

If granulation is studied at those places where it is produced best, it becomes evident that the grains have a wide variety of shapes, but are more or less spherical.

This form is usually much better defined for the smaller elements. In the case of a multiple granulation, with a more or less irregular shape, it is apparent that such a granulation is an agglomeration of the smaller spherical elements.

Even when the granulation is less exact, and the elements appear to be deformed, it still seems that the elements were originally spherical in shape, and that the forces acting on the bodies have caused some modification in their shapes.

Thus the normal form of the granular elements of the photosphere resembles a sphere, and the irregular shapes still seem to be related to that form, whether the granulation in question is made up of smaller bodies, or whether it has been deformed to some extent by external forces acting on the medium in which it is situated. A very important fact may follow from these circumstances; the evidence of the large variety of shapes of the granular elements points to the likelihood of their being made up of a highly mobile medium, which is very susceptible to change by external forces. Both the liquid and gaseous states possess such properties, but in relation to factors which will be mentioned in due course, it seems that for granulations to exist the state of the medium must be very similar to that of atmospheric clouds. In other words, we consider that they are bodies made up of fine particles of solid or liquid matter floating in a gaseous medium.

Origin of Granulations

If the solar layer which comprises the photosphere was in a state of perfect equilibrium, then, by the principle of fluidity, the result would be the formation of a continuous envelope around the solar nucleus. The granular elements would mingle together and the radiance of the Sun would be completely uniform. The ascending currents of gas do not permit that state of perfect equilibrium, however, since the currents break up and divide each fluid layer into a large number of points as they pass through: this leads to the formation of the granulations as elements of the photospheric envelope. These granulations tend to assume spherical shapes by virtue of the gravity of their constituent parts: thus it becomes apparent that the globular form is typical of a state of relative rather than absolute equilibrium, the photospheric medium having been unable to maintain a continuous layer and having split up into elements, each of which takes on its own equilibrium shape. But even this state of equilibrium of the various parts is comparatively rare, because at various places the gaseous currents have to carry the granular elements more or less along with them, and the globular equilibrium shape can be altered to something quite unrecognizable when the induced motion becomes more violent.

These movements, which constantly disturb that part of the gaseous layer where the photospheric elements float, have areas of variable effect. The solar surface is thus divided into regions of comparative calm and of activity: hence the photospheric net-work. Moreover, even in the calmer regions, the motion of the photospheric medium does not permit the granular elements to spread out into a uniform layer, with the result that there is a fairly considerable concentration of grains just below the surface and, since the medium in which these elements are suspended has a high absorption, there are marked differences in the brightness of their photographic images.

Thus a preliminary study of the latest photographs shows that we must modify considerably our ideas about the photosphere; the facts which they present to us lead to a fairly simple explanation of the photospheric elements, and of the transformations which they undergo due to the various forces acting on them.

We can pursue this result to the conclusion that, since there is a relatively small number of very brilliant granules on the photographic images, the luminous power of the Sun stems mainly from that small number of dots on its surface. That is, if the solar surface were covered entirely by granular elements of the most radiant kind, then its luminous power would be ten or twenty

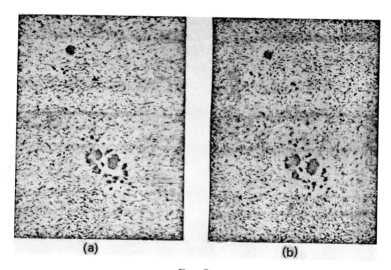

Fig. 7.

times greater, according to a preliminary estimate to which we will return.

Finally, there is the major problem on which these investigations throw new light, namely the much debated question of the variation in the Sun's luminous power. It is evident that the spots can no longer be regarded as the principal elements in the variations which the Sun displays, and that in future it will be necessary to take account of the number and luminous power of the granular elements, which are very important for this.

Apart from sunspots, one of the most important uses of photography is in the study of other motions and transformations on the surface of the Sun.

That such a study is a fairly simple matter can be seen by a brief examination of the images of these phenomena. The pictures show the granular elements, their shapes and grouping, and all other features of the surface, provided they are at least a quarter of a second of arc across. The main requirement of the apparatus is that it has a high-precision reticle, since it is important that the position located by the wires is better defined than the size of the details which are to be observed. So far, it has not been possible to produce an instrument capable of such delicate adjustment, but the time will soon come when one is available.

By the examination of successive photographs, taken at brief intervals, some new phenomena have already been revealed. I have already established that photospheric granules have quite short life-times, that they are subject to rapid transformation, and that they are activated particularly by the ascending streams of hydrogen. In this connection the discovery of the photospheric network has been a great help. It is apparent that the shape and position of the polygons undergo rapid changes.

In order that our readers may appreciate both the existence of the photospheric network and the above-mentioned changes, we have reproduced by the photoglyph method two images of an average portion of the Sun's surface, taken 50 minutes apart (see FIG. 7). We have deliberately chosen a region which includes a spot because its appearance makes it easier to align and to carry out measurements which show the movements under consideration.

From the photographic point of view, however, it must be appreciated that the reproduction of the spot has been sacrificed for reproduction of the granulation. This is because the time of action of light in the case of a spot is much greater than that for the details of the photospheric surface. In this case, the spot simply acts as a point of reference for the study of the movement of the granules.

The fact that these reproductions are photoglyphs means that the human hand plays no part in their production. The original negatives were used to make positives on a plate of gelatine bichromate on lead. When the plate has been washed off and then hardened it becomes possible to etch the lead plate, which can be used to print the image on paper. The firm of Lemercier & Co. have carried out this delicate work very satisfactorily.

These are only the preliminary results of what promises to be an extremely rewarding technique.

Photography cannot only be applied to the study of the Sun. At the moment, Mr. Huggins is obtaining stellar spectra which show very significant results. Mr. Lockyer is examining the solar spectrum with particular reference to metallic spectra, and he has reached the conclusion that an entirely new range of metals are present in the Sun.

Photographic experiments are under way for the production of stellar celestial maps. Mr. Rutherfurd, who took the beautiful lunar photographs which caused such a sensation recently, is considering this important question. Mr. Gould, in South America, has already succeeded in photographing stars of the eleventh magnitude. In France, Mr. Ed. Becquerel, the first person to obtain a pure ultraviolet spectrum, and after him Messrs. Mascart and Cornu, have used photography to study the most refractive parts of the solar spectrum and to produce scientific maps. In this manner, photography has achieved considerable importance in astronomy. If, as we hope, the observatory at Meudon becomes well endowed, then these rather neglected but highly relevant fields will be the subject of future research.

Finally, we return to the subject of the Sun, and see if the facts we have just established might permit a more complete synthesis concerning its nature and origin. As science progresses, the nebular and stellar origin of the Sun appears more and more likely. Hydrogen is the principal element in the composition of the nebulae, and of the major stars. Now hydrogen plays a large part in the composition of the Sun; the hydrogen gas comes from the depths of the photosphere, rises through a cloudy layer of dust or metallic particles, which it disturbs and activates, carries them to the surface and causes them to shine efficiently for us. The gas then continues to rise and forms the prominences which extend into the coronal atmosphere, an ultimate atmosphere, a medium of transition between the heavy metallic vapours of the photosphere and the celestial regions. As for the photospheric clouds (our granulations), it is probable that for them hydrogen plays a role similar to that of the atmospheric currents which support terrestrial clouds and prevent them from falling onto the surface of the Earth. The radiating quality of the Sun depends exclusively on the photospheric clouds, and their survival has been well provided for. Here I must mention a very important fact, which was clearly laid down by Mr. Faye in his excellent theory of the Sun.

Is it not obvious that these photospheric clouds, if they were firmly attached to the solar surface, would soon be consumed by their own radiation?

This radiation, however, is simply indicating the state of condensation of the clouds; it makes their constituent parts heavier, and probably causes the formation of larger solid or liquid particles. These particles fall towards the centre of the Sun by virtue of their weight, vaporize, and are brought to the surface again by currents of hydrogen, and so the cycle continues indefinitely.

A necessary condition for the occurrence of such phenomena is that, at a certain depth the Sun must be warmer than at its surface. Indeed, the longer I continue my studies the more I am led to believe in the existence of a solar nucleus or reservoir of heat, its function being to support the photosphere. We have seen, in fact, how the photospheric clouds disperse when they are moved towards the bottom of the spots. But above all it is for reasons of a *philosophical nature* that I accept this truth. In fact an increasing amount of evidence points to the nebular origin of the Sun. How can one conceive of a nebula, having incandescent gases, which is able to create suns with cold nuclei? The condensation can only cause an increase in heat, and cannot diminish it.

Moreover, the theory of functions must surely be of increasing relevance in astronomy. Is not the solar system an organism of a kind where each part has its separate function? That of the Sun is to distribute its heat and light to the surrounding planets by radiation. The planets are destined to be the theatre of life. To populate the Sun would be to create havoc in the harmony of the universe.

Thus the Sun can radiate its health-giving rays to its satellites for as long as it is required to do so. A law which is based on the fundamental properties of matter states that the entire mass of the Sun can be called upon to support its radiating power and that it would need to exhaust that vast reservoir of power before reaching the state which exists at the surface. We can rest assured, therefore, that although our own Sun is not the youngest and brightest of all the suns and stars, nevertheless it has the potential to satisfy the most ambitious dreams of mankind.

10

On the Distribution
of the Solar Spots in
Latitude since the Beginning
of the Year 1854†

R. C. CARRINGTON

I HAD fully hoped that by the time when the Society would meet
again in the month of November I should have been able to
present the members with a tolerably complete discussion of the
series of solar spot observations which I have now kept up at
Redhill for nearly five years; but a family loss by the hand of death
has for the last three months deprived me of the leisure I had
counted on, and compels me to confine the present paper to the
publication of a single feature which the assembling and comparison of results has brought to light. It will be found to be
another and instructive instance of the regular irregularity and the
irregular regularity which, in the present state of our knowledge,
appear to characterise the solar phenomena.

In the briefest form of statement, the result is, that throughout
the two years preceding the minimum of frequency in February
1856, the spots were confined to an equatoreal belt, and in no
instance passed the limits of 20° of latitude N. or S.; and that
shortly after this epoch, whether connected with it or not, this
equatoreal series appears to have become extinct, and in seeming
contradiction to the precept, *Natura non agit per saltum*, two new
belts of disturbance abruptly commenced, the limits of which in
both hemispheres may be roughly set at between 20° and 40°,
with exceptions in favour of the old equatoreal region. The

† *Mon. Not. Roy. Astr. Soc.* **19**, 1–3, 1858.

tendency at the present time appears to be to contraction of the parallels.

In a mean solar day the sun rotates through an angle of 14° 11′ very nearly, and the heliographical longitudes of spots as compared with that of the centre of the disk, increase by the synodic quantity corresponding. If an observer plots down on each day the positions reduced to noon of the spots recorded on the 180° viewed, and lays the successive sheets so as to superpose the positions of the same spot, his plotting sheets will advance daily to the left by 13° 12′ nearly. And if he then agrees to adopt one figure of a spot as typical, to be laid down on a single continuous sheet, he will obtain a result such as I have deduced by computation (with the exception that for convenience my map is reversed left for right), and such as might be supposed to be produced by imagining the sun to be the central cylinder of one of Applegarth's printing-machines. In this state the graphical result would be of too extreme dimensions in length for the eye to take it in as a whole; and, accordingly, in the map lithographed the scale in longitude is reduced in the proportion of 72 to 1, as compared with the scale in latitude, and the delineations of spots compressed to mere vertical lines. The uncertainty of the exact period of rotation affects the result only in this way, that if the true period be somewhat longer than that adopted, the vertical lines which indicate the commencement of successive rotations, should be drawn somewhat wider apart, and vice versa; a matter of no importance, obviously, for our present purpose.

The variation of the limiting parallels being established, the inquirer next desires to turn back to such past records as may throw light on their changes and their relation to epochs. In this respect I would call attention to a short, but very condensed and important paper, communicated to me by Dr. C. H. F. Peters, formerly of Naples, and lately of Albany, and published, I believe,† by the American Association for the Advancement of Science. Dr. Peters observed the sun with an object and purpose

† I have by me only the sheets of the volume containing this particular paper, sent by post.

very similar to my own, from September 1845 to October 1846 inclusive, and obtained 813 places of 286 spots, which he subsequently reduced with a skill and exactness in which I place great confidence. He has given a plate exhibiting the gross distribution in latitude during the period of his labours, in which I find the limits laid down as 40° N. and 30° S. with a desolate region from 8° N. to 5° S., and with, on the whole, a preponderance of action in the north hemisphere. We know now that at that time the sun was passing from a period of minimum activity to a maximum, as now. The distribution is very similar to that which now holds, excepting that at the present time there appears an excess of activity in the south. The records of Dr. Peters are not all which leisure and research may make available in this branch of the history of solar action, there are those of Soemmering and others; nevertheless, it is difficult to express the degree of regret which a student of the sun feels when evidence such as the present meets him of the state of maturity his subject might have attained ere this, had not the opportunities of two centuries been neglected, by his predecessors condemning the research as one of idle curiosity, fit matter for a University thesis, but below the level of Philosophy.

The most cursory consideration will show that success in educing such conclusions in the case of the sun depends mainly on the continuity of the labours of the observer. The conclusion which an observer would have arrived at from a discussion of the observations made during the years 1854 and 1855 would have been exceedingly imperfect, though apparently borne out by a tolerably extensive experience; and the conclusion which I draw from the four years' results now accumulated is, that our knowledge of the sun's action is but fragmentary, and that the publication of speculations on the nature of his spots would be a very precarious venture.

I am very anxious to know what the magnetic observers have to produce, corresponding or not corresponding with the results of my map; and I would take the opportunity of remarking that the question of the correspondency of the solar and magnetic disturbance phenomena is in the curious and imperfect state of a correspondency established in the aggregate, but not for particulars.

I shall shortly offer some conclusions on the independent movement of spots, and on the divergence of neighbouring nuclei, a very singular and marked action, in the detection of which I find, however, that Dr. Peters has anticipated me.

11

Spectroscopic Observations of the Sun†

J. Norman Lockyer

The two most recent theories dealing with the physical constitution of the sun are due to M. Faye and to Messrs. De la Rue, Balfour Stewart, and Loewy. The chief point of difference in these two theories is the explanation given by each of the phenomena of sun-spots.

Thus, according to M. Faye,‡ the interior of the sun is a nebulous gaseous mass of feeble radiating-power, at a temperature of dissociation; the photosphere is, on the other hand, of a high radiating-power, and at a temperature sufficiently low to permit of chemical action. In a sun-spot we see the interior nebulous mass through an opening in the photosphere, caused by an upward current, and the sun-spot is black, by reason of the feeble radiating-power of the nebulous mass.

In the theory held by Messrs. De la Rue, Stewart, and Loewy,§ the appearances connected with sun-spots are referred to the effects, cooling and absorptive, of an inrush, or descending current, of the sun's atmosphere, which is known to be colder than the photosphere.

In June 1865 I communicated to the Royal Astronomical Society‖ some observations (referred to by the authors last named) which had led me independently to the same conclusion as the one announced by them. The observations indicated that,

† *Proc. Roy. Soc.* **15,** 256–8, 1866.
‡ *Comptes Rendus* **60,** 89–138, abstracted in *The Reader*, 4 February 1865.
§ *Researches on Solar Physics.* Printed for private circulation, Taylor and Francis, 1865.
‖ *Monthly Notices Roy. Ast. Soc.* **25,** 237.

instead of a spot being caused by an *upward* current, it is caused by a *downward* one, and that the results, or, at all events, the concomitants of the downward current are a dimming and possible vaporization of the cloud-masses carried down. I was led to hold that the current had a downward direction by the fact that one of the cloud-masses observed passed in succession, in the space of about two hours, through the various orders of brightness exhibited by *faculae*, general surface, and *penumbrae*.

On March 4th of the present year I commenced a spectroscopic observation of sun-spots, with a view of endeavouring to test the two rival theories, and especially of following up the observations before alluded to.

The method I adopted was to apply a direct-vision spectroscope to my 6¼-inch equatoreal (by Messrs. Cooke and Sons) at some distance outside the eyepiece, with its axis coincident with the axis of the telescope prolonged. In front of the slit of the spectroscope was placed a screen on which the image of the sun was received; in this screen there was also a fine slit corresponding to that of the spectroscope.

By this method it is possible to observe at one time the spectra of the umbra of a spot and of the adjoining photosphere or penumbra; unfortunately, however, favourable conditions of spot (i.e. as to size, position on the disk, and absence "of cloudy stratum"), atmosphere, and instrument are rarely coincident. The conditions were by no means all I could have desired when my first observations were made; and, owing to the recent absence of spots, I have had no opportunities of repeating my observations. Hence I should have hesitated still longer to lay them before the Royal Society had not M. Faye again recently called attention to the subject.

On turning the telescope and spectrum-apparatus, driven by clock-work, on to the sun at the date mentioned, in such a manner that the centre of the umbra of the small spot then visible fell on the middle of the slit in the screen, which, like the corresponding one in the spectroscope, was longer than the diameter of the umbra, the solar spectrum was observed in the field of view of the spectroscope with its central position (corresponding to the diameter of the umbra falling on the slit) greatly enfeebled in brilliancy.

All the absorption bands, however, visible in the spectrum of the photosphere, above and below, were visible in the spectrum of the spot; they, moreover, appeared thicker where they crossed the spot-spectrum.

I was unable to detect the slightest indication of any bright bands, although the spectrum was sufficiently feeble, I think, to have rendered them unmistakeably visible had there been any.

Should these observations be confirmed by observations of a larger spot free from "cloudy stratum," it will follow, not only that the phenomena presented by a sun-spot are not due to radiation from such a source as that indicated by M. Faye, but that we have in this absorption-hypothesis a complete or partial solution of the problem which has withstood so many attacks.

The dispersive power of the spectroscope employed was not sufficient to enable me to determine whether the decreased brilliancy of the spot-spectrum was due in any measure to a greater number of bands of absorption, nor could I prove whether the thickness of the bands in the spot-spectrum, as compared with their thickness in the photosphere-spectrum, was real or apparent only.†

On these points, among others, I shall hope, if permitted, to lay the results of future observations before the Royal Society. Seeing that spectrum-analysis has already been applied to the stars with such success, it is not too much to think that an attentive and *detailed* spectroscopic examination of the sun's surface may bring us much knowledge bearing on the physical constitution of that luminary. For instance, if the theory of absorption be true, we may suppose that in a deep spot rays might be absorbed which would escape absorption in the higher strata of the atmosphere; hence also the darkness of a line may depend somewhat on the depth of the absorbing atmosphere. May not also some of the

† Irradiation would cause bands of the same thickness to appear thinnest in the more brilliant spectrum.

variable lines visible in the solar spectrum be due to absorption in the region of spots? and may not the spectroscope afford us evidence of the existence of the "red flames" which total eclipses have revealed to us in the sun's atmosphere; although they escape all other methods of observation at other times? and if so, may we not learn something from this of the recent outburst of the star in Corona?

12

Spectroscopic Notes†

C. A. Young

For the past few months I have been examining the spectra of sun-spots with great care, and with an instrument of high dispersion. The spectroscope employed consists of a comet-seeker of five inches aperture and about forty-eight inches focal length, used both as collimator and view-telescope after Littrow's method, the slit and diagonal eye-piece being as close together as it is convenient to place them. A small spot of black paper, about three tenths of an inch in diameter, is cemented to the centre of the object-glass (as suggested by my colleague, Professor Brackett, in a note published in the *American Journal of Science*, July 1882), and entirely destroys the internal reflections, which would otherwise most seriously interfere with vision.

The dispersion is obtained by one of Professor Rowland's magnificent gratings on a speculum-metal plane, with a ruled surface three and one-half inches by five, 14,000 lines to the inch. The slit and eye-piece of the telescope are so placed that the line joining them is parallel to the lines of the ruling, An instrument of this sort is incomparably more convenient than one in which the collimator and view-telescope are separate, though of course, on account of the inclination of the visual rays to the axis of the object-glass, there is a little aberration, and the *maximissimum* of definition is not quite reached. There is no difficulty, however, in seeing easily the duplicity of the b's, E_1, and other similar tests with the instrument thus arranged. The spectroscope is mounted upon a strong plank, stiffly braced, and this is attached by powerful

† *Phil. Mag.* **16** (Ser. 5), 460–3, 1883.

ring-clamps to the tail-piece of the 23-inch equatoreal of the Halsted observatory, so that the image of the sun falls directly upon the slit.

The detailed examination of the spot-spectra has been thus far confined mainly to a few limited regions in the neighbourhood of C, D, and *b*.

With the high dispersive power employed, the widening and "winging" of the heavier lines of the spectrum is not well seen, not nearly so well as with a single-prism spectroscope. All diffuse shadings disappear much in the same way as the naked-eye markings on the moon's face vanish in a powerful telescope—to be replaced by others more minute but not less interesting. In a few spots however, the broadening of the D's and the reversal and occasional "lumping" of C has been noticeable even with this high dispersion. But the most striking result is that, in certain regions the spectrum of the spot-nucleus, instead of appearing as a mere continuous shade, crossed here and there by markings dark and light, is resolved into a countless number of lines, exceedingly fine and closely packed, interrupted frequently between E and F (and occasionally below E) by lines as bright as the spectrum outside the spot. These bright lines, so far as the eye can determine, may be either real lines *superposed*, or merely vacancies left in the shading of fine dark lines, since they are not sensibly brighter than the ordinary background of the surrounding spectrum.

The darker and more intense the spot, the more distinctly the fine lines come out, both the bright and the dark; and so far as I have been able to make out yet, there is no difference as regards these fine lines between one spot and another. I have never yet seen any evidence of displacement in them due to motion, no "lumpiness" nor want of smoothness in them.

When seeing is at the best, and everything favourable, close attention enables one to trace nearly all these lines out beyond the spot and its penumbra. But they are so exceedingly faint on the sun's general surface that usually they cannot be detected outside the spot-spectrum. This resolution of the spot-spectrum into a congeries of fine lines is most easily made out in the green and blue. Near D, and below it, it is much more difficult to see; and I am not even quite sure that this structure still exists in the regions around C and below it. Here, in the red, even with the highest dispersion and under the most favourable circumstances of vision, the spot-spectrum appears simply as a continuous shade, crossed here and there by widened and darkened lines, which, however, are very few and far between as compared with the number of such lines in the higher regions. Of course the resolution of the spot-spectrum into lines tends to indicate that the absorption which darkens the centre of a sun-spot is produced, not by granules of solid or liquid matter, but by matter in the gaseous form; and it becomes interesting to inquire what substances are capable of producing such a spectrum, and under what conditions. As to the fineness and number of the lines, it may be noted that in the region included between b_1 and b_2 the single lines appear to be each about half as wide as the components of b_3, and are separated by an interval about one third as great. The whole number between b_1 and b_4 must be over a hundred, though they are of course very difficult to count with accuracy. They are a little wider in the middle of the spot-spectrum, in fact spindle-shaped, running out into extremely fine threads where they pass into the penumbra, and in my instrument they seem to be a little more hardly and sharply defined on the upper (more-refrangible) edge than on the lower.

The bright lines, of which there are six between b_1 and b_4, are generally about as wide as their interspace between the components of b_3. They are sharply defined at both edges, and no brighter at the centre than at the edge—a fact which rather bears in favour of the idea that they are merely interruptions in the dark-line series, and not really superposed bright lines. Just above b_4 (at λ 5162.3) there is a very conspicuous one, which is also noticeable enough in the ordinary solar spectrum. Attention has indeed been frequently called to it long since by other observers. Below E these bright lines are rare. Higher up in the spectrum, between F and G, they become very numerous.

I have also made a considerable number of observations upon prominences with the nine and one-half inch equatoreal and its own spectroscope. There have been lately numerous very fine exhibitions, especially in connection with the spots. The number of lines reversed in the spectrum of the chromosphere has at times been very great, far exceeding the number observed and catalogued in 1872; but I have not been able to detect a single *new* one below C, though the two mentioned in my catalogue have been seen almost continuously. On two occasions (July 31st and August 1st) a new line above H (λ 3884 $\pm$ 2) was conspicuously visible for an hour or two each time, during a specially vigorous eruption of the prominences associated with the great spot, which was then just passing off the limb. This line was seen easily without the aid of any fluorescent eye-piece; and I am satisfied that on a photographic plate it would have been more brilliant than either H or K. I could not determine its position within one or two units on account of the difficulty of identifying the numberless fine lines around it.

With the widened slit it showed clearly the form of the lower part of the prominence, but not the upper. It was almost precisely imitated by two new lines at λ 4092 and 4026; and the catalogue-lines 4077 and 3990 resembled it also. On the other hand, *h*, H, and K showed the higher parts of the prominence as well as the lower, while the lines at 4045 and 3970 were exceedingly fine and smooth, without knottiness or structure.

On August 1st, at 2^h 58^m local time ($= 7^h$ 57^m Greenwich time), the intensity of the chromosphere-spectrum was very remarkable, the bright lines more vivid and numerous than I remember ever to have seen them before. Between this time and 3.12 a prominence was shot up in fragments of flame to an elevation of over 120,000 miles. It will be interesting to learn whether any corresponding magnetic twitch appears on the magnetometer records.

13

Description of a Singular Appearance
seen in the Sun on September 1, 1859†

R. C. CARRINGTON

WHILE engaged in the forenoon of Thursday, Sept. 1, in taking my customary observation of the forms and positions of the solar spots, an appearance was witnessed which I believe to be exceedingly rare. The image of the sun's disk was, as usual with me, projected on to a plate of glass coated with distemper of a pale straw colour, and at a distance and under a power which presented a picture of about 11 inches diameter. I had secured diagrams of all the groups and detached spots, and was engaged at the time in counting from a chronometer and recording the contacts of the spots with the cross-wires used in the observation, when within the area of the great north group (the size of which had previously excited general remark), two patches of intensely bright and white light broke out, in the positions indicated in fig. 1 by the letters A and B, and of the forms of the spaces left white. My first impression was that by some chance a ray of light had penetrated a hole in the screen attached to the object-glass, by which the general image is thrown into shade, for the brilliancy was fully equal to that of direct sun-light; but, by at once interrupting the current observation, and causing the image to move by turning the R.A. handle, I saw I was an unprepared witness of a very different affair. I thereupon noted down the time by the chronometer, and seeing the outburst to be very rapidly on the increase, and being somewhat flurried by the surprise, I hastily ran to call some one to witness the exhibition with me, and on returning

† *Monthly Notices of the Roy. Astron. Soc.* **20**, 13–15, 1860.

within 60 seconds, was mortified to find that it was already much changed and enfeebled. Very shortly afterwards the last trace was gone, and although I maintained a strict watch for nearly an hour, no recurrence took place. The last traces were at C and D, the patches having travelled considerably from their first position

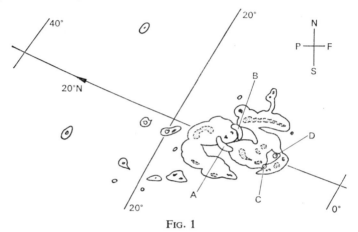

FIG. 1

and vanishing as two rapidly fading dots of white light. The instant of the first outburst was not 15 seconds different from 11ʰ 18ᵐ Greenwich mean time, and 11ʰ 23ᵐ was taken for the time of disappearance. In this lapse of 5 minutes, the two patches of light traversed a space of about 35,000 miles, as may be seen by the diagram, which is given exactly on a scale of 12 inches to the sun's diameter. On this scale the section of the earth will be very nearly equal in area to that of the detached spot situated most to the north in the diagram, and the section of *Jupiter* would about cover the area of the larger group, without including the outlying portions. It was impossible, on first witnessing an appearance so similar to a sudden conflagration, not to expect a considerable result in the way of alteration of the details of the group in which it occurred; and I was certainly surprised, on referring to the sketch which I had carefully and satisfactorily (and I may add fortunately) finished before the occurrence, at finding myself unable to recognize any change whatever as having taken place. The impression left upon me is, that the phenomenon took place at an elevation considerably above the general surface of the sun, and, accordingly, altogether above and over the great group in which it was seen projected. Both in figure and position the patches of light seemed entirely independent of the configuration of the great spot, and of its parts, whether nucleus or umbra. The customary observation was shortly resumed, and the diagram engraved, as well as the larger drawing exhibited at the Meeting on Nov. 11, was deduced from an exact reduction of the recorded times.

It has been very gratifying to me to learn that our friend Mr. Hodgson chanced to be observing the sun at his house at Highgate on the same day, and to hear that he was a witness of what he also considered a very remarkable phenomenon. I have carefully avoided exchanging any information with that gentleman, that any value which the accounts may possess may be increased by their entire independence.

[Mr. Carrington exhibited at the November Meeting of the Society a complete diagram of the disk of the sun at the time, and copies of the photographic records of the variations of the three magnetic elements, as obtained at Kew, and pointed out that a moderate but very marked disturbance took place at about $11^h 20^m$ A.M., Sept. 1st, of short duration; and that towards four hours after midnight there commenced a great magnetic storm, which subsequent accounts established to have been as considerable in the southern as in the northern hemisphere. While the contemporary occurrence may deserve noting, he would not have it supposed that he even leans towards hastily connecting them. "One swallow does not make a summer."]

14

On a Curious
Appearance seen in the Sun†

R. HODGSON

WHILE observing a group of solar spots on the 1st September, I was suddenly surprised at the appearance of a very brilliant star of light, much brighter than the sun's surface, most dazzling to the protected eye, illuminating the upper edges of the adjacent spots and streaks, not unlike in effect the edging of the clouds at sunset; the rays extended in all directions; and the centre might be compared to the dazzling brilliancy of the bright star α-*Lyrae* when seen in a large telescope with low power. It lasted for some five minutes, and disappeared instantaneously about 11.25 a.m. Telescope used, an equatoreal refractor 6 inches aperture, carried by clockwork; power, a single convex lens, 100, with a pale neutral-tint sunglass; the whole aperture was used with a diagonal reflector.

The phenomenon was of too short duration to admit of a micrometrical drawing, but an eye-sketch was taken, from which an enlarged diagram has been made; and from a photograph taken at Kew the previous day, the size of the group appears to have been about $2^m 8^s$, or (say) 60,000 miles.

The magnetic instruments at Kew were simultaneously disturbed to a great extent.

† *Monthly Notices of the Roy. Astron. Soc.* **20**, 15–16, 1860.

15

The Effective Temperature of the Sun†

W. E. Wilson

In March 1894 Dr. G. Johnston Stoney communicated to the Society a memoir by myself and Mr. P. L. Gray, entitled "Experimental Investigations on the Effective Temperature of the Sun," which was published in the *Phil. Trans.*, A, vol. 185 (1894). In these investigations the method we adopted was as follows. A beam of sunlight was sent horizontally into the laboratory by means of a Stoney single-mirror heliostat. The mirror was an optical plane of unsilvered glass, and the beam was directed into one aperture (A) of a differential Boys' radio-micrometer. The other aperture (B) received the radiation from a strip of platinum, which could be raised to any desired temperature by an electric current supplied by a battery of accumulators. The temperature of this strip was at any moment determined by its linear expansion, the instrument being previously calibrated by melting on it minute fragments of AgCl and of pure gold, as in Joly's meldometer. In front of the aperture (B) of the radio-micrometer was placed a stop with a circular hole of 5·57 mm, and the distance of this hole from the receiving surface of the thermo-couple was 60·2 mm. This gave for the angle subtended by a diameter of the aperture at the receiving surface 5°·301. Knowing then (i) the ratio which the angular diameter of this circular aperture bears to that of the sun, (ii) the temperature of the platinum strip at the moment that the radio-micrometer is balanced, (iii) the amount of the sun's radiation lost by reflection from the heliostat mirror and also by absorption in the earth's atmosphere, it is possible on any assump-

† *Proc. Roy. Soc.* **69**, 312–20, 1901.

tion with regard to the law connecting radiation with temperature, to determine the effective temperature of the sun. A series of very accordant observations were made in this way, the mean of which gave 6200° C as the effective solar temperature.

For the details of the apparatus and the complete method of reduction of the observations, the original memoir in the *Phil. Trans.* may be referred to.

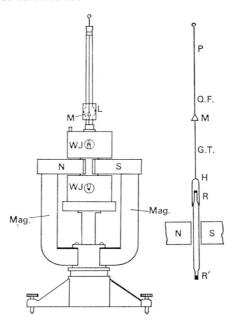

In order to protect the incandescent strip from draughts of air it was covered with a water-jacket of gilded brass. This was provided with a circular hole in one of its longer sides, through which its radiation could reach the aperture of the radio-micrometer. The internal walls of this water-jacket being highly polished, it has occurred to me, since the publication of the memoir referred to, that possibly some of the radiation from distant parts of the platinum strip may have been reflected backwards and forwards from the polished walls and the strip itself, ultimately escaping through the aperture and reaching the radio-micrometer, thus increasing the amount of radiation which should have reached it directly from the strip alone.

In order to test this surmise I first took a number of readings at known temperatures with the walls of the water-jacket polished as before. I next smoked the surface of the walls well, and found that the amount of radiation coming from the aperture was then sensibly reduced. It is also possible that changes in the condition of the surface of the platinum strip may affect its emissivity, and in fact it is very doubtful whether it is possible to determine with any degree of accuracy what the emissivity of bright platinum is, relatively to lamp black. In the original memoir we took Rosetti's estimate of 35 per cent as the most probable value for this quantity, but as our former estimate of the solar temperature depends greatly on this factor, to which so much uncertainty attaches, I thought it would be a distinct advance to abolish entirely the platinum strip as a source of radiation, and to substitute in its place a uniformly heated enclosure which would radiate as an absolutely "black body."

In 1895 Mr. Lanchester pointed out to me that such an enclosure would be a theoretically perfect radiator; while Lummer, Paschen and others have shown that the law connecting temperature and radiation from such an enclosure confirms in a remarkable manner Stefan's law of radiation, viz, $R = aT^4$.[†] Since therefore the results of several independent investigations corroborate this law, I have felt justified in applying it to the results of my observations.

On consideration it seemed that the most convenient form of radiator would be a long tube closed at one end, and uniformly heated in a gas furnace. Accordingly a porcelain tube, 2 feet in length and 1 inch internal diameter, was fitted into a Fletcher gas-tube furnace. This was afterwards changed for an iron tube, which

[†] Stefan, *Sitzber. Akad. Wien* **79** (2), 391, 1879; Boltzmann, *Wied. Ann.* **22**, 1884; M. Planck, *Drude Ann.* **1**, No. 1, 1900; Paschen, *Wied Ann* **58, 60,** 1896, 1897; Lummer and Pringsheim, *Wied. Ann.* **63**, 395, 1897.

was employed in the observations on September 30th, given below.

A plug of asbestos was inserted in the tube at about 10 inches from the end farthest from the radio-micrometer, and resting against this plug was the end of a Callendar platinum-resistance thermometer. This was connected with one of Professor Callendar's electric recorders, so that during an experiment the temperature of the tube was registered continuously on the paper wrapped round the drum of the instrument. In front of the open end of the tube, and between it and the radio-micrometer, was placed a large brass water-screen, through which a copious supply of water passed. In front of the aperture (B) of the radio-micrometer this screen was provided with a rectangular aperture. One side of this aperture was formed by a slide moved by a micrometer screw reading to 0·01 mm. By this means the area of this aperture at any time could be measured with precision. Its fixed sides were 5 mm apart, and as the movable side had a range of 5 mm, the maximum area of the aperture was 25 mm. The distance (d) of this aperture from the surface of the thermocouple was 66·3 mm.

To make an observation the tube was heated to as high a temperature as the furnace was capable of, and when a steady temperature had been obtained, the amount of radiant heat coming from the interior of the hot tube and passing into the aperture (B) of the radio-micrometer was adjusted by the micrometer screw until a balance was obtained with the radiation coming from the sun through the aperture (A).

If ρ is the angular semi-diameter of the sun, its radiation outside of our atmosphere is $K\pi \sin^2 \rho$, where K is a constant depending on the sun's temperature.

Again, if a be the height of the slit through which the radiation from the hot tube reaches the radio-micrometer, and β its width, the radiation may with sufficient accuracy be expressed by $Ka\beta/d^2$. Assuming Stefan's law, the radiation of the sun outside our atmosphere is $a\theta^4\pi \sin^2\rho$, θ being the effective temperature of the sun.

The percentage transmitted is p, therefore the radiation before reflection from the heliostat is $\dfrac{ap\theta^4}{100} \pi \sin^2 \rho$.

At reflection q per cent is transmitted, therefore the radiation received by the radio-micrometer is

$$a\frac{pq\theta^4}{(100)^2}\,\pi\sin^2\!\rho \tag{1}$$

Also the radiation received from the hot tube is

$$a'(T^4-T_0{}^4)\,\frac{\alpha\beta}{d^2} \tag{2}$$

We need not inquire what is the absorptive power of the thermo-junction, provided that we are justified in assuming that lamp-blacked surfaces absorb the radiation from the hot tube as freely as that from the sun, or that the constants in these expressions (1) and (2) for the radiation may be taken to be equal.

On balancing, these expressions must be equal, and therefore

$$\frac{pq\theta^4}{(100)^2}\,\pi\,\sin^2\!\rho=(T^4-T_0{}^4)\,\frac{\alpha\beta}{d^2}$$

$$=T^4\left[1-\left(\frac{T_0}{T}\right)^4\right]\frac{\alpha\beta}{d^2} \tag{3}$$

But $\left(\dfrac{T_0{}^4}{T}\right)$ may be neglected, hence we have finally

$$\theta^4=\frac{\alpha}{\pi d^2}\,\frac{\beta}{\sin^2\!\rho}\,\frac{100}{p}\,\frac{100}{q}\,T^4$$

or

$$\theta=\sqrt[4]{\frac{10000\alpha}{\pi d^2}}\,\sqrt[4]{\frac{\beta}{pq\,\sin^2\!\rho}}\,T$$

$$=\frac{0{\cdot}13806}{\sqrt{\sin\rho}}\,\sqrt[4]{\frac{\beta}{pq}}\,T.$$

The mean value of $\dfrac{0{\cdot}13806}{\sqrt{\sin\rho}}$ is [1·30413].

Therefore $$\theta=1{\cdot}30413\,\sqrt[4]{\frac{\beta}{pq}}\,T \tag{4}$$

After a series of observations had been made, the furnace and tube were raised so that the radiation of the latter then passed into the aperture (A), on which the sunlight had previously fallen, while the beam of sunlight was now directed so as to be upon (B), and in this position a second series of observations was taken. The geometrical mean of the result of the two groups gives the effective temperature of the Sun, the effect of any difference in the sensitiveness of the thermo-junctions disappearing in the geometrical mean.

Observations were made in the manner described above on August 19th and September 30th, 1901, and reduced by means of equation (4).

Rosetti's determination of the amount of the terrestrial atmospheric absorption was used. It may be well, however, to give the results obtained by using other estimates of this quantity. Taking Langley's transmission coefficient when the sun is in the zenith as 59 per cent, compared to Rosetti's 71 per cent, the temperature would be multiplied by $\sqrt[4]{(71/59)}$ and thus become $5773 \times 1 \cdot 054$, which is $6085°$ absolute. And, as in the previous memoir, to make the case general, if any later investigation shows the zenith transmission coefficient to be x per cent, the effective absolute temperature becomes

$$5773° \times \sqrt[4]{(71/x)}.$$

It may also be of interest to see what effect is produced if absorption in the atmosphere of the sun itself is taken into account. First, considering the falling off in radiation from the central to the peripheral parts of the sun's disc, we may deduce that, if the absorption were everywhere equal to that at the centre, the radiation would be multiplied by $4/3$ and the temperature would become

$$5773° \times \sqrt[4]{(4/3)} = 5773 \times 1 \cdot 074 = 6201°.$$

Secondly, assuming Wilson and Rambaut's[†] result for the total loss due to absorption in the solar atmosphere as equal to one-third, our estimate of the temperature would have to be multiplied by $\sqrt[4]{(3/2)}$, and we get finally

$$6201° \times \sqrt[4]{(3/2)} = 6201° \times 1 \cdot 107 = 6863° \text{ absolute} = 6590° \text{ C.}$$

I wish to express my thanks to Dr. Rambaut for some valuable suggestions during the progress of the work.

[†] The Absorption of Heat in the Solar Atmosphere, *Proc. Royal Irish Academy* **2**, No. 2, 1892.

16

Spectroscopic Observations of the Sun. III†

J. Norman Lockyer

Since my second paper under the above title was communicated to the Royal Society, the weather has been unfavourable to observatory work to an almost unprecedented degree; and, as a consequence, the number of observations I have been enabled to make during the last four months is very much smaller than I had hoped it would be.

Fortunately, however, the time has not been wholly lost in consequence of the weather; for, by the kindness of Dr. Frankland, I have been able in the interim to familiarize myself at the Royal College of Chemistry with the spectra of gases and vapours under previously untried conditions, and, in addition to the results already communicated to the Royal Society by Dr. Frankland and myself, the experience I have gained at the College of Chemistry has guided me greatly in my observations at the telescope.

In my former paper it was stated that a diligent search after the known third line of hydrogen in the spectrum of the chromosphere had not met with success. When, however, Dr. Frankland and myself had determined that the pressure in the chromosphere even was small, and that the widening out of the hydrogen lines was due in the main, if not entirely, to pressure, I determined to seek for it again under better atmospheric conditions; and succeeded after some failures. The position of this third line is at 2796 of Kirchhoff's scale. It is generally excessively faint,

† *Proc. Roy. Soc.* **17**, 350–6, 1869.

and much more care is required to see it than is necessary in the case of the other lines; the least haze in the sky puts it out altogether.

Hence, then, with the exception of the bright yellow line, the observed spectra of the prominences and of the chromosphere correspond exactly with the spectrum of hydrogen under different conditions of pressure—a fact not only important in itself, but as pointing to what may be hoped for in the future.

With regard to the yellow line which Dr. Frankland and myself have stated may possibly be due to the radiation of a great thickness of hydrogen, it became a matter of importance to determine whether, like the red and green lines (C and F), it could be seen extending on to the limb. I have not observed this: it has always in my instrument appeared as a very fine sharp line resting absolutely on the solar spectrum, and never encroaching on it.

Dr. Frankland and myself have pointed out that, although the chromosphere and the prominences give out the spectrum of hydrogen, it does not follow that they are composed merely of that substance: supposing others to be mixed up with hydrogen, we might presume that they would be indicated by their selective absorption near the sun's limb. In this case the spectrum of the limb would contain additional Fraunhofer lines. I have pursued this investigation to some extent, with, at present, negative results; but I find that special instrumental appliances are necessary to settle the question, and these are now being constructed.

If we assume, as already suggested by Dr. Frankland and myself that no other extensive atmosphere besides the chromosphere overlies the photosphere, the darkening of the limb being due to the general absorption of the chromosphere, it will follow:

I. That an additional selective absorption near the limb is extremely probable.

II. That the hydrogen Fraunhofer lines indicating the absorption of the outer shell of the chromosphere will vary somewhat in thickness: this I find to be the case to a certain extent.

III. That it is not probable that the prominences will be visible on the sun's disk.

In connexion with the probable chromospheric darkening of the limb, an observation of a spot on February 20th is of importance. The spot observed was near the limb, and the absorption was much greater than anything I had seen before; so great, in fact, was the *general* absorption, that the several lines could only be distinguished with difficulty, except in the very brightest region. I ascribe this to the greater length of the absorbing medium in the spot itself in the line of sight, when the spot is observed near the limb, than when it is observed in the centre of the disk—another indication of the great general absorbing power of a comparatively thin layer, on rays passing through it obliquely.

I now come to the selective absorption in a spot. I have commenced a map of the spot-spectrum, which, however, will require some time to complete. In the interim, I may state that the result of my work up to the present time in this direction has been to add magnesium and barium to the material (sodium) to which I referred in my paper in 1866, No. I. of the present series; and I no longer regard a spot simply as a cavity, but as a place in which principally the vapours of sodium, barium, and magnesium (owing to a downrush) occupy a lower position than they do ordinarily in the photosphere.

I do not make this assertion merely on the strength of the lines observed to be thickest in the spot-spectrum, but also upon the following observations on the chromosphere made on the 21st and 28th ultimo.

On both these days the brilliancy of the F line taught me that something unusual was going on; so I swept along the spectrum to see if any materials were being injected into the chromosphere.

On the 21st I caught a trace of magnesium; but it was late in the day, and I was compelled to cease observing by houses hiding the sun.

On the 28th I was more fortunate. If anything, the evidences of intense action were stronger than on the 21st, and after one glance at the F line I turned at once to the magnesium lines. I saw them appearing short and faint at the base of the chromosphere. My work on the spots led me to imagine that I should find sodium-vapour associated with the magnesium; and on turning from *b* to D I found this to be the case. I afterwards reversed barium in the same way. The spectrum of the chromosphere seemed to be full of lines, and I do not think the three substances I have named accounted for all of them. The observation was one of excessive delicacy, as the lines were short and *very thin*. The prominence was a small one, about twice the usual height of the chromosphere; but the hydrogen lines towered high above those due to the newly injected materials. The lines of magnesium extended perhaps one-sixth of the height of the F line, barium a little less, and sodium least of all.

We have, then, the following facts:

    I. The lines of sodium, magnesium, and barium, when observed in a spot, are thicker than their usual Fraunhofer lines.

    II. The lines of sodium, magnesium, and barium, when observed in the chromosphere, are thinner than their usual Fraunhofer lines.

A series of experiments bearing upon these observations is now in progress at the College of Chemistry, and will form the subject of a communication from Dr. Frankland and myself. I may at once, however, remark that we have here additional evidence of a fact I asserted in 1865 on telescopic evidence—the fact, namely, that a spot is the seat of a downrush, a downrush to a region, as we now know, where the selective absorption of the upper strata is different from what it would be (and, indeed, is elsewhere) at a higher level.

Messrs. De la Rue, Stewart, and Loewy, who brought forward the theory of a downrush about the same time as my observations were made in 1865, at once suggested as one advantage of this explanation that all the gradations of darkness, from the faculae to the central umbra, are thus supposed to be due to the same cause, namely, the presence to a greater or less extent of a relatively cooler absorbing atmosphere. This I think is now spectroscopically established; we have, in fact, two causes for the darkening of a spot:

I. The general absorption of the chromosphere, thicker here than elsewhere, as the spot is a cavity.

II. The greater selective absorption of the lower sodium, barium, magnesium stratum, the surface of its last layer being below the ordinary level.

Messrs. De La Rue, Stewart, and Loewy also suggested, in their 'Researches on Solar Physics,' that if the photosphere of the sun be the plane of condensation of gaseous matter, the plane may be found to be subject to periodical elevations and depressions, and that at the epoch of minimum sun-spot-frequency the plane might be uplifted very high in the solar atmosphere, so that there was comparatively little cold absorbing atmosphere above it, and therefore great difficulty in forming a spot.

This suggestion is one of great value; and, as I pointed out in my previous paper, its accuracy can fortunately now be tested. It may happen, however, that in similar periodical fluctuations the chromosphere may be carried up and down with the photosphere; and I have already evidence that possibly such a state of things may have occurred since 1860, for I do not find the C and F Fraunhofer lines of the same relative thickness as they were in that year.† I am waiting to make observations with the large Steinheil spectroscope before I consider this question settled. But the well-known great thickness of the F line in Sirius and other stars will point out the excessive importance of such observations as a method of ascertaining not only the physical constitution, but the actual pressures of the outer limits of stellar atmospheres, and of the same atmosphere at different epochs. And when other spectra have been studied as we have now studied hydrogen, additional means of continuing similar researches will be at our command; indeed a somewhat careful examination of the spectra of the different classes of stars, as defined by Father Secchi, leads me to believe that several broad conclusions are not far to seek; and I hope soon to lay them before the Royal Society.

† I have learnt, after handing this paper in to the Royal Society, that in Ångström's Map the C and F lines are nearly of the same breadth: this I had gathered from observations made with my own spectroscope.

For some time past I have been engaged in endeavouring to obtain a sight of the prominences, by using a very rapidly oscillating slit; but although I believe this method will eventually succeed, the spectroscope I employ does not allow me to apply it under sufficiently good conditions, and I am not at present satisfied with the results I have obtained.

Hearing, however, from Mr. De La Rue, on February 27th, that Mr. Huggins had succeeded in anticipating me by using absorbing media and a wide slit (the description forwarded to me is short and vague), it immediately struck me, as possibly it has struck Mr. Huggins, that the wide slit is quite sufficient without any absorptive media; and during the last few days I have been perfectly enchanted with the sight which my spectroscope has revealed to me. The solar and atmospheric spectra being hidden, and the image of the wide slit alone being visible, the telescope or slit is moved slowly, and the strange shadow-forms flit past. Here one is reminded, by the fleecy, infinitely delicate cloud-films, of an English hedgerow with luxuriant elms; here of a densely intertwined tropical forest, the intimately interwoven branches threading in all directions, the prominences generally expanding as they mount upwards, and changing slowly, indeed almost imperceptibly. By this method the smallest details of the prominences and of the chromosphere itself are rendered perfectly visible and easy of observation.

Addendum

Since the foregoing paper was written, I have had, thanks to the somewhat better weather, some favourable opportunities for continuing two of the lines of research more especially alluded to in it; I refer to the method I had adopted for viewing the prominences, and to the injection of sodium, magnesium, etc. into the chromosphere.

With regard to seeing the prominences, I find that, when the sky is free from haze, the views I obtain of them are so perfect that I have not thought it worth while to remount the oscillating slit. I am, however, collecting red and green and violet glass, of the required absorptions, to construct a rapidly revolving wheel, in which the percentages of light of each colour may be regulated. In this way I think it possible that we may in time be able to see the prominences as they really are seen in an eclipse, with the additional advantage that we shall be able to see the sun at the same time, and test the connexion or otherwise between the prominences and the surface-phenomena.

Although I find it generally best for sketching-purposes to have the open slit in a radial direction, I have lately placed it at a tangent to the limb, in order to study the general outline of the chromosphere, which in a previous communication I stated to be pretty uniform, while M. Janssen has characterized it as *à niveau fort inégal et tourmenté*. My opinion is now that perhaps the mean of these two descriptions is, as usual, nearer the truth, unless the surface changes its character to a large extent from time to time. I find, too, that in different parts the outline varies: here it is undulating and billowy; there it is ragged to a degree, flames, as it were, darting out of the general surface, and forming a ragged, fleecy, interwoven outline, which in places is nearly even for some distance, and, like the billowy surface, becomes excessively uneven in the neighbourhood of a prominence.

According to my present limited experience of these exquisitely beautiful solar appendages, it is generally possible to see the whole of their structure; but sometimes they are of such dimensions along the line of sight that they appear to be much denser than usual; and as there is no longer under these circumstances any background to the central portion, only the details of the margins can be observed, in addition to the varying brightnesses.

Moreover it does not at all follow that the largest prominences are those in which the intensest action, or the most rapid change, is going on—the action as visible to us being generally confined to the regions just in, or above, the chromosphere, the changes arising from violent uprush or rapid dissipation, the uprush and dissipation representing the birth and death of a prominence. As a rule, the attachment to the chromosphere is narrow and is not often single; higher up, the stems, so to speak, intertwine, and the prominence expands and soars upward until it is lost in delicate filaments, which are carried away in floating masses.

Since last October, up to the time of trying the method of using the open slit, I had obtained evidence of considerable changes in the prominences from day to day. With the open slit it is at once evident that changes on the small scale are continually going on; it was only on the 14th inst. that I observed any change at all comparable in magnitude and rapidity to those already observed by M. Janssen.

About $9^h 45^m$ on that day, with a tangential slit I observed a fine dense prominence near the sun's equator, on the eastern limb. I tried to sketch it with the slit in this direction; but its border was so full of detail, and the atmospheric conditions were so unfavourable, that I gave up the attempt in despair. I turned the instrument round 90° and narrowed the slit, and my attention was at once taken by the F line; a single look at it taught me that an injection into the chromosphere and intense action were taking place. These phenomena I will refer to subsequently.

At $10^h 50^m$, when the action was slackening, I opened the slit; I saw at once that the dense appearance had all disappeared, and cloud-like filaments had taken its place. The first sketch, embracing an irregular prominence with a long perfectly straight one, which I called A, was finished at $11^h 5^m$, the height of the prominence being $1' 5''$, or about 27,000 miles. I left the Observatory for a few minutes; and on returning, at $11^h 15^m$, I was astonished to find that part of the prominence A had entirely disappeared; not even the slightest rack appeared in its place: whether it was entirely dissipated, or whether parts of it had been wafted towards the other part, I do not know, although I think the latter explanation the more probable one, as the other part had increased.

We now come to the other attendant phenomena. First, as to the F line. In my second paper, under the above title, I stated that the F line widens as the sun is approached, and that sometimes the bright line seems to extend on to the sun itself, sometimes on one side of the F line, sometimes on the other.

Dr. Frankland and myself have pointed out, as a result of a long series of experiments, that the widening out is due to pressure, and apparently not to temperature *per se*; the F line near the vacuum-point is thin, and it widens out on both sides (I do not say to the same extent) as the pressure is increased. Now, in the absence of any disturbing cause, it would appear that when the wider line shows itself on the sun on one side of the F line, it should at the same time show itself on the other; this, however, *it does not always do.* I have now additional evidence to adduce on this point, and this time in the prominence line itself, off the sun. In the prominence to which I have referred, the F bright line underwent the most strange contortions, as if there were some disturbing cause which varied the refrangibility of the hydrogen-line under certain conditions and pressures.

The D line of hydrogen (?) also once bore a similar appearance.

Secondly, as to the other phenomena which accompanied this strange behaviour of the F line, and were apparently the cause of it.

In the same field of view with F, I recognized the barium-line at 1989·5 of Kirchhoff's scale.

Passing on, the magnesium-lines and the enclosed nickel-iron-line were visible in the chromosphere. The magnesium was projected higher into the chromosphere than the barium, and the nickel or iron was projected higher than the magnesium. I carefully examined whether the other iron-lines were visible in the spectrum of the chromosphere; they were not.

I also searched for the stronger barium-lines in the brighter portion of the spectrum; but I did not find them, probably owing to the feeble elevation of the barium-vapour above the general level of the photosphere, which made the observation in this region a very delicate one.

I detected another chromosphere-line very near the iron-line at 1569·5 (on the east side of it).

The sodium-lines were also visible.

Unfortunately clouds prevented my continuing these interesting observations; but the action was evidently toning down.

Here, then, we have an uprush of
  Barium,
  Magnesium,
  ? Nickel,
  and an unknown substance
from the photosphere into the chromosphere, and with the uprush
a *dense* prominence; accompanying the uprush we have changes
of an enormous magnitude in the prominence; and as the uprush
ceases the prominence melts away.

As stated in the former part of this paper, the barium- and
magnesium-lines were thinner than the corresponding Fraunhofer
lines. In connexion with this subject, I beg to be allowed to state
that I have commenced a careful comparison of Kirchhoff's map
with the recently published one of Ångström. From what I have
already seen, I believe other important conclusions, in addition
to that before alluded to, may be derived from this comparison;
but I hesitate to say more at present, as I have not yet been able
to compare Ångström's maps with the sun itself, or to examine
the angular diameters of the sun registered at Greenwich during
the present century.

On the 14th inst. I also succeeded in detecting the hydrogen-line
in the extreme violet in the spectrum of the chromosphere.

17

Preliminary Note of Researches on Gaseous Spectra in relation to the Physical Constitution of the Sun†

Edward Frankland and J. Norman Lockyer

1. For some time past we have been engaged in a careful examination of the spectra of several gases and vapours under varying conditions of pressure and temperature, with a view to throw light upon the discoveries recently made bearing upon the physical constitution of the sun.

Although the investigations are by no means yet completed, we consider it desirable to lay at once before the Royal Society several broad conclusions at which we have already arrived.

It will be recollected that one of us in a recent communication to the Royal Society pointed out the following facts:

i. That there is a continuous envelope round the sun, and that in the spectrum of this envelope (which has been named for accuracy of description the "chromosphere") the hydrogen line in the green corresponding with Fraunhofer's line F takes the form of an arrowhead, and widens from the upper to the lower surface of the chromosphere.

ii. That ordinarily in a prominence the F line is nearly of the same thickness as the C line.

iii. That sometimes in a prominence the F line is exceedingly brilliant, and widens out so as to present a bulbous appearance above the chromosphere.

iv. That the F line in the chromosphere, and also the C line,

† *Proc. Roy. Soc.* **17,** 288–91, 1869.

extend on to the spectrum of the subjacent regions and re-reverse the Fraunhofer lines.

v. That there is a line near D visible in the spectrum of the chromosphere to which there is no corresponding Fraunhofer line.

vi. That there are many bright lines visible in the ordinary solar spectrum near the sun's edge.

vii. That a new line sometimes makes its appearance in the chromosphere.

2. It became obviously, then, of primary importance:

i. To study the hydrogen spectrum very carefully under varying conditions, with the view of detecting whether or not there existed a line in the orange, and

ii. To determine the cause to which the thickening of the F line is due.

We have altogether failed to detect any line in the hydrogen spectrum in the place indicated, i.e. near the line D; but we have not yet completed all the experiments we had proposed to ourselves.

With regard to the thickening of the F line, we may remark that, in the paper by MM. Plücker and Hittorf, to which reference was made in the communication before alluded to, the phenomena of the expansion of the spectral lines of hydrogen are fully stated, but the cause of the phenomena is left undetermined.

We have convinced ourselves that this widening out is due to pressure, and not appreciably, if at all, to temperature *per se*.

3. Having determined, then, that the phenomena presented by the F line were phenomena depending upon and indicating varying pressures, we were in a position to determine the atmospheric pressure operating in a prominence, in which the red and green lines are nearly of equal width, and in the chromosphere, through which the green line gradually expands as the sun is approached.†

With regard to the higher prominences, we have ample evidence that the gaseous medium of which they are composed exists in a

† Will not this enable us ultimately to determine the temperature?

condition of *excessive* tenuity, and that at the lower surface of the chromosphere itself the pressure is very far below the pressure of the earth's atmosphere.

The bulbous appearance of the F line before referred to may be taken to indicate violent convective currents or local generations of heat, the condition of the chromosphere being doubtless one of the most intense action.

4. We will now return for one moment to the hydrogen spectrum. We have already stated that certain proposed experiments have not been carried out. We have postponed them in consequence of a further consideration of the fact that the bright line near D has apparently no representative among the Fraunhofer lines. This fact implies that, assuming the line to be a hydrogen line, the selective absorption of the chromosphere is insufficient to reverse the spectrum.

It is to be remembered that the stratum of incandescent gas which is pierced by the line of sight along the sun's limb, the radiation from which stratum gives us the spectrum of the chromosphere, is very great compared with the radial thickness of the chromosphere itself; it would amount to something under 200,000 miles close to the limb.

Although there is another possible explanation of the non-reversal of the D line, we reserve our remarks on the subject (with which the visibility of the prominences on the sun's disk is connected) until further experiments and observations have been made.

5. We believe that the determination of the above-mentioned facts leads us necessarily to several important modifications of the received theory of the physical constitution of our central luminary—the theory we owe to Kirchhoff, who based it upon his examination of the solar spectrum. According to this hypothesis, the photosphere itself is either solid or liquid, and it is surrounded by an atmosphere composed of gases and the vapours of the substances incandescent in the photosphere.

We find, however, instead of this compound atmosphere, one which gives us nearly, or at all events mainly the spectrum of hydrogen; (it is not, however, composed necessarily of hydrogen alone; and this point is engaging our special attention;) and the tenuity of this incandescent atmosphere is such that it is extremely improbable that any considerable atmosphere, such as the corona has been imagined to indicate, lies outside it,—a view strengthened by the fact that the chromosphere bright lines present no appearance of absorption, and that its physical conditions are not statical.

With regard to the photosphere itself, so far from being either a solid surface or a liquid ocean, that it is cloudy or gaseous or both follows both from our observations and experiments. The separate prior observations of both of us have shown:

i. That a gaseous condition of the photosphere is quite consistent with its continuous spectrum. The possibility of this condition has also been suggested by Messrs. De La Rue, Stewart, and Loewy.

ii. That the spectrum of the photosphere contains bright lines when the limb is observed, these bright lines indicating probably an outer shell of the photosphere of a gaseous nature.

iii. That a sun-spot is a region of greater absorption.

iv. That occasionally photospheric matter appears to be injected into the chromosphere.

May not these facts indicate that the absorption to which the reversal of the spectrum and the Fraunhofer lines are due takes place in the photosphere itself or extremely near to it, instead of in an extensive outer absorbing atmosphere? And is not this conclusion strengthened by the consideration that otherwise the newly discovered bright lines in the solar spectrum itself should be themselves reversed on Kirchhoff's theory? this, however, is not the case. We do not forget that the selective radiation of the chromosphere does not necessarily indicate the whole of its possible selective absorption; but our experiments lead us to believe that, were any considerable quantity of metallic vapours present, their bright spectra would not be entirely invisible in all strata of the chromosphere.

18

The New Spectrum†

S. P. LANGLEY

THE writer (at the concluding meeting of the National Academy of Sciences on April 18) remarked on the disadvantages in the matter of interest of the work of the physicist, which he was about to show them, to that of the biologist, which was concerned with the ever absorbing problem of life. He had, however, something which seemed to him of interest, even in this respect, to speak of, for it included some indications he believed to be new, pointing the way to future knowledge of the connexion of terrestrial life with that physical creator of all life, the sun.

He had to present to the Academy a book embodying the labour of twenty years, though at this late hour he could scarcely more than show the volume with a mention of the leading captions of its subject. What he had to say then would be understood as only a sort of introductory description of the contents of the work in question, which was Volume I of the *Annals of the Astrophysical Observatory of the Smithsonian Institution.*

In illustration of a principal feature of this book, the Academy saw before them on the wall an extended solar spectrum, only a small portion of the beginning of which, on the left, was the visible spectrum known to Sir Isaac Newton. This was the familiar visible coloured spectrum which we all have seen and know something of, even if our special studies are in other fields.

It is chiefly this visible part which has been hitherto the seat of prolonged spectroscopic investigation, from a little beyond the

† *Phil. Mag.* (6th Ser.) **2**, 119–30, 1901. Abstract of a paper read before the National Academy of Sciences at its Washington meeting, April 18, 1901.

violet, at a wave-length of somewhat less than 0.4^{μ}, down to the extreme red, which is generally considered to terminate at the almost invisible line A, whose wave-length is 0.76^{μ}. On the scale of the actual wave-length of light, then, where the unit of measurement (1^{μ}) is one one-thousandth of a millimetre, the length of the visible spectrum is 0.36^{μ}.

The undue importance which this visible region has assumed, not only in the eyes of the public, but in the work of the spectroscopist, is easily intelligible, being due primarily to the evident fact that we all possess, as a gift from nature, a wonderful instrument for noting the sun's energy in this part, and in this part only.

While, then, this part alone can be seen by all, yet the idea of its undue importance is also owing to the circumstance that the operation of the ordinary prism gives an immensely extended linear depiction of the really small amount of energy in this visible part. There is also a region beyond the violet, most insignificant in energy and invisible to the eye, and the association of this linear extension due to the prism, with the accident that the salts of silver used in photography are extraordinarily sensitive to these short wave-length rays, so that they can depict them even through the most extreme enfeeblement of the energy involved in producing them, also makes this part have undue prominence. This action of the prism and of the photograph is local, then, and peculiar to the short wave-lengths; and owing to it, all but special students of the subject are, as a rule, under a wholly erroneous impression of the relative importance of what is visible and what is not. The spectrum has really no positive dimension, being extended at one end or the other according to the use of the prism or grating employed in producing it. Perhaps the only fair measurement for displaying a linear representation of the energy would be that of a special scheme, which the writer had proposed, in which the energy is everywhere the same;† but this presentation is unusual and would not be generally intelligible without explanation.

† *Amer. Journ. Science* **27** (Ser. 3), 169, 1884.

The map before us will be intelligible when it is stated that it is, as to the infra-red, an exact representation of that part of the spectrum given by a rock-salt prism. The visible and ultra-violet spectrum given here is not exact, for the reason that it would take nearly a hundred *feet* of map to depict it on the prismatic scale, though this is caused by but a small fraction of the sun's energy; so monstrous is the exaggeration due to the dispersion of the prism.

Looking, then, at the map: First, in the spectrum on the left and beyond 0.4^μ is the ultra-violet region, in fact almost invisibly small, but which in most photographs shows almost a *hundred times larger than the whole infra-red.* It really contains much less than one hundredth part of the total solar energy which exists. Beyond is the visible spectrum, containing perhaps one fifth of the solar energy.

As the writer has elsewhere said, "the amount of energy in any region of the spectrum, such as that in any colour, or between any two specified limits, is a definite quantity, fixed by facts, which are independent of our choice, such as the nature of the radiant body or the absorption which the ray has undergone. Beyond this Nature has no law which must govern us."

Everything in the linear presentation, then, depends on the scale adopted. In other words, if we have the lengths proportionable to the energies, the familiar prismatic representation enormously exaggerates the importance of the visible, and still more of the ultra-violet region, and similarly the grating spectrum exaggerates that of the infra-red region. Now he had given, on the map before them, and through the whole infra-red, the exact rock-salt prismatic spectrum, but for the purpose of obtaining a length which represented (though insufficiently) that of the visible spectrum, he had laid the latter down on the *average* dispersion in the infra-red, which was perhaps as fair a plan as could be taken for showing the approximate relation of the two fields of energy in an intelligible way, though it gave the visible energy too small.

Let us recall, then, at the risk of iteration, that in spite of the familiar extended photographic spectra of the hundreds of lines shown in the ultra-violet, and in those of the coloured spectrum, it is not here that the real creative energy of the sun is to be studied, but elsewhere, on the right of the drawing, in the infra-red. Looking to the spectrum as thus delineated, next to the invisibly small and weak ultra-violet, comes the visible or Newtonian spectrum, which is here somewhat insufficiently shown, and on the right extends the great invisible spectrum in which four-fifths of the solar energies are now known to exist.

Of this immense invisible region nothing was known until the year 1800,† when Sir William Herschel found heat there with the thermometer.

After that little was done‡ (except an ingenious experiment by Sir John Herschel§ to show that the heat was not continuous) till the first drawing of the energy curve by Lamansky,‖ in 1871, which, on account of its great importance in the history of the subject, is given on the map. It consists of the energy curves of the visible spectrum, and beyond it, on the right (and in illustration of what has just been said it will be seen how relatively small these latter appear), of three depressions indicating lapses of heat in the infra-red. It is almost impossible to tell what these lapses are meant for, without a scale of some kind (which he does not furnish), but they probably indicate something, going down to near a wave-length of 1^{μ}. It is obvious that the detail is of the

† *Phil. Trans.* **90**, 284, 1800.

‡ It should, however, be mentioned that an important paper by Draper (*Lond. Ed. & Dublin Phil. Mag.*, May 1843) was published in 1843, in which he appears to claim the discovery of the group here called $\rho\sigma\tau$ and which is now known to have a wave-length of less than 1^{μ}. (Its true wave-length was not determined till much later.) Later, Fizeau seems to have found further irregularities of this heat as long ago as 1847, and of its location, obtaining his wave-lengths by means of interference-bands. His instrumental processes, though correct in theory, were not exact in practice; and yet it seems pretty clear that he obtained some sort of recognition of a something indicating heat, as far down as the great region immediately above Ω on our present charts. Mouton (*Comptes Rendus*, 1879) confirmed this observation of Fizeau's and contrived to get at least an approximate wave-length of the point where the spectrum (to him) ended, at about $1\cdot8^{\mu}$.

§ *Phil. Trans.* **130**, 1, 1840.

‖ *Monatsberichte der k. Akademie der Wissenschaften zu Berlin*, December 1871.

very crudest, and yet this drawing of Lamansky's was remarkable as the first drawing of the energy spectrum. It attracted general attention, and was the immediate cause of the writer's taking up his researches in this direction.

It seems proper to state here that the true wave-lengths were at that time most imperfectly known, but that in 1884, and later in 1885,[†] they were completely determined by the writer as far as the end of what he has called "the new spectrum" at a wave-length of $5 \cdot 3^{\mu}$.

The upper portion of the infra-red is quite accessible to photography, and the next important publication in this direction was that of Captain (now Sir William) Abney,[‡] which gave the photographic spectrum down to about $1 \cdot 1^{\mu}$, much beyond which photography has never mapped since.

From the time of seeing Lamansky's drawing, the writer had grown interested in this work, but found the thermopile, the instrument of his predecessors, and the most delicate then known to science, insufficient in the feeble heat of the grating spectrum, and about 1880 he had invented the bolometer[§] and was using it in that year for these researches. This may perhaps seem the place to speak of this instrument, though with the later developments which have made it what it is today, it has grown to something very different from what it was then.

It has, in fact, since found very general acceptance among physicists, especially since it has lately reached a degree of accuracy, as well as of delicacy, which would have appeared impossible to the inventor himself in its early days.

It may be considered in several relations, but notably as to three: (1) Its sensitiveness to small amounts of heat; (2) the accuracy of measurement of those small amounts: and (3) the accuracy of its measurements of the position of the source of heat.

As to the first, it is well known that the principle of the instrument depends on the forming of a Wheatstone bridge, by the

† *Amer. Journ. Science*, March 1884 and August 1886.
‡ *Phil. Trans.* **171**, 653, 1880.
§ Actinic balance, *Amer. Journ. Science* **21** (Ser. 3), 187, 1881.

means of two strips of platinum or other metal, of narrow width and still more limited thickness, one of which only is exposed to the radiation. In some bolometers in use, for instance, the strip is a tenth of a millimetre, or one two-hundredth-and-fiftieth of an inch in width; and yet it is to be described as only a kind of tape, since its thickness is less than a tenth of this.

The use of the instrument is then based on the well-known fact that the heating of an ordinary metallic conductor increases its resistance, and this law is found to hold good in quantities so small that they approach the physically infinitesimal. In the actual bolometers, for instance, the two arms of a Wheatstone bridge are formed of two strips of platinum, side by side, one of which is exposed to the heat and the other sheltered. The warming of the exposed one increases its resistance and causes a deflexion of the galvanometer.

It was considered to be remarkable twenty years ago that a change of temperature of one ten-thousandth of a degree centigrade could be registered; it is believed at present that with the consecutive improvements of the original instrument and others, including those which Mr. Abbot, of the Smithsonian Institution Observatory, has lately introduced into its attendant galvanometer, less than one one-hundred-millionth of a degree in the change of temperature of the strip can be registered. This indicates the sensitiveness of the instrument to heat.

As to the second relation, some measures have been made of the steadiest light source obtainable. With ordinary photometric measures of its intensity one might expect a probable error of about 1 per cent. The error with the bolometer was insensible by any means that could be applied to test it. It is at any rate less than two one-hundredths of 1 per cent. If we imagine an absolutely invisible spectrum, in which there nevertheless are interruptions of energy similar to those which the eye shows us in the visible, then the bolometer, whose sensitive strip passes over a dark line in the spectrum, visible or invisible (since what is darkness to the eye is cold to it), gives a deflexion on the side of cold, and in the warmer interval between two lines a deflexion on the side of heat;

these deflexions being proportionate to the cause, within the degree of accuracy just stated.

The third quality, the accuracy of its measures of position, is better seen by a comparison and a statement, for if we look back to the indications of the lower part of Lamansky's drawing we may see that at least a considerable fraction of a degree of error must exist there in such a vague delineation. Now, in contrast with this early record, the bolometer has been brought to grope in the dark and to thus feel the presence of narrow Fraunhofer-like lines by their cooler temperature alone, with an error of the order of that in refined astronomical measurement; that is to say, the probable error, in a mean of six observations of the relative position of one of these invisible lines, is less than one second of arc; a statement which the astronomer, perhaps, who knows what an illusive thing a second of arc is, can best appreciate.

The results of the writer's labours with the bolometer in the years 1880 and 1881, and in part of his expedition in the latter year to Mount Whitney, were given at the Southampton meeting of the British Association for the Advancement of Science in 1882.† During these two years very many thousand galvanometer readings were taken, by a most tryingly slow process, to give the twenty or more interruptions shown at that time, below the limit of $1·1^\mu$ of Abney's photographs. The bolometer has been called an eye which sees in the dark, but at that time the "eye" was not fairly open, and having then not been brought to its present rapidity of use, the early results were attained only by such un-limited repetition, and almost infinite patience was needed till what was inaccurate was eliminated.

Several hundreds at least of galvanometer readings were then taken to establish the place of each of the above twenty lines during the two years when they were being hunted for, and this patience so far found its reward that they have never required any material alteration since, but only additions such as the writer can now give. The part below $1·1^\mu$ he then presented (at the Southampton meeting of the British Association) as having been mapped

† *Report British Association*, 1882. *Nature* **26**, 1882.

for the first time. Mouton had two years before obtained crude indications of heat as far as $1·8^{\mu}$, and Abney had, as stated, obtained relatively complete photographs of the upper infra-red extending to about this point ($1·1^{\mu}$).

The writer had already determined for the first time by the bolometer, at Allegheny and on Mount Whitney, the wave-lengths of some much remoter regions, including, in part, the region then first discovered by him and here called "the new spectrum", and was able to state that the terminal ray of the solar spectrum, whose presence had then been certainly felt by the bolometer, had a wave-length of about $2·8^{\mu}$, or nearly two octaves below the "great A" of Fraunhofer.

He stated in this communication of 1882 that the galvanometer then responded readily to changes of temperature in the bolometer strip of much less than one ten-thousandth of a degree centigrade (as has just been said, it now responds to changes of less than one one-hundred-millionth), and he added: "Since it is one and the same solar energy, whose manifestations are called 'light' or 'heat' according to the medium which interprets them, what is 'light' to the eye is 'heat' to the bolometer, and what is seen as a dark line by the eye is felt as a cold line by the sentient instrument. Accordingly, if lines analogous to the dark 'Fraunhofer' lines exist in this invisible region, they will appear (if I may so speak) to the bolometer as cold bands, and this hair-like strip of platinum is moved along in the invisible part of the spectrum till the galvanometer indicates the all but infinitesimal change of temperature caused by its contact with such a 'cold band'. The whole work, it will be seen, is necessarily very slow; it is, in fact, a long groping in the dark and it demands extreme patience."

At that time it may be said to have been shown that these interruptions were due to the existence of something like dark lines or bands, resembling what are known as the Fraunhofer lines in the upper spectrum; but, apart from what the writer had done, no one then surmised how far this spectrum extended nor, perhaps, what these explorations really meant. They may be compared to actual journeys into this dark continent, if it may be so called, which extended so far beyond those of previous explorers that the determination of positions by the writer, corresponding somewhat to longitudes determined by the terrestrial explorer in a new country, was, by those who had not been so far but had conceived an inadequate idea of the extent of the region, treated as erroneous and impossible.

A necessary limit to the farthest infra-red was in 1880 supposed to exist near the wave-length 1^μ. Doctor John Draper,† for instance, announced in other terms that the extreme end of the invisible spectrum might, from theoretical considerations, be probably estimated at something less than the wave-length of 1^μ, whence it followed that the above value of $1\cdot8^\mu$ was impossible, and, still more, that of $2\cdot8^\mu$. If, in this connexion, we revert to our map, where the visible spectrum has an extent in wave-lengths of $0\cdot36^\mu$, then, on that same scale, the length of the entire possible spectrum visible and invisible, was fixed by Draper at the point there shown near the band $\rho\sigma\tau$. In still other words, according to him the very end of any spectrum at all would be about 3 on a scale in which the visible spectrum was 1. Doctor Draper's authority was deservedly respected, and this citation of his remarks is made only to show the view then entertained by eminent men of science.

Now, the writer had proved by actual measurement that it extended far beyond this point, and had announced, as the result of experiment, that it extended at any rate to about three times the utmost length then assigned from theoretical reasons, by Draper, founded on the then universally accepted formula of Cauchy, which was later discredited by the direct experimental evidence given of its falsity by the bolometer.

The bolometer, which is wholly independent of light as a sensation and notes it only as a manifestation of energy, first lays down the spectrum by curves of energy from which the linear spectrum is in turn derived. Two such curves taken at different times are given to show the agreement.

There must now be explained, however briefly, the way in

† *Proc. Amer. Acad.* **16,** 233, 1880.

which these energy curves, which are the basis of all, have actually been produced here.

In making the map of the energy curves, it should be remembered that when an invisible band or line is suspected, its presence is revealed by the change of temperature in the bolometer strips affecting the needle of the galvanometer, causing this needle to swing this way or that; let us suppose to the left if from cold and the right if from heat. The writer's first method was to have one person to note the exposure, another to note the extent of the deflexion, and a third to note the part of the spectrum in which it occurred. For reasons into which he does not enter, this old plan was, as he has already said, tedious in the extreme and required, as has been said, hundreds of observations to fix with appropriate accuracy the position in wave-length of one invisible line. It has been stated that only about twenty such lines had been mapped out in nearly two years of assiduous work prior to 1881, and if a thousand such lines existed, it was apparent that fifty years would be required to denote them.

The writer then devised a second apparatus to be used in connexion with the bolometer. This apparatus was simple in theory, though it has taken a dozen years to make it work well in practice, but it is working at last, and with this the maps in this volume of the 'Annals' and that before us have been chiefly made. It is almost entirely automatic, and, as it is now used, a thousand inflections can be delineated in a single hour, much better than this could have been done in the half-century of work just referred to.

Briefly, the method is this: A great rock-salt prism (for a glass one would not transmit these lower rays nor could they easily be detected in the overlapping spectra of the grating) is obtained of such purity and accuracy of figure, and so well sheltered from moisture, that its clearness and its indications compare favourably, even in the visible spectrum, with those of the most perfect prism of glass, with the additional advantage that it is permeable to the extreme infra-red rays in question. This prism rests on a large azimuth circle turned by clockwork of the extremest precision, which causes the spectrum to move slowly along, and in one minute of time, for example, to move exactly one minute of arc of its length before the strip of the bolometer, bringing this successively in contact with one invisible line and another. Since what is blackness to the eye is cold to the bolometer, the contact of the black lines chills the strip and increases the electric current. The bolometer is connected by a cable with the galvanometer, whose consequent swing to the right or the left is photographically registered on a plate which the same clockwork causes to move synchronously and uniformly up or down by exactly one centimetre of space for the corresponding minute. By this means the energy-curve of an invisible region, which directly is wholly inaccessible to photography, is photographed upon the plate.

Let it be noted that whatever the relation of the movement of the spectrum to that of the plate is (and different ones might be adopted), it is absolutely synchronous—at least to such a degree that an error in the position of one of these invisible lines can be determined, as has been stated, with the order of precision of the astronomical measurement of visible things.

The results were before them in the energy-curves and the linear infra-red spectrum, containing over seven hundred invisible lines. This is more than the number of visible ones in Kirchhoff and Bunsen's charts. The position of each line is fixed from a mean of at least six independent determinations with the accuracy stated above.

The reader will perhaps gather a clearer idea of this action if he imagines the map before him hung up at right angles to its actual position, so that a rise in the energy-curve given would be seen to correspond to a deflexion to the right, and a fall, to one to the left; for in this way the deflexions were written down on the moving photographic plate from which this print has been made. The writer was now speaking of the refinements of the most recent practice; but there was something in this retrospect of the instrument's early use which brought up a personal reminiscence which he asked the Academy to indulge him in alluding to.

This was that of one day in 1881, nearly twenty years ago, when being near the summit of Mount Whitney, in the Sierra

Nevadas, at an altitude of 12,000 feet, he there, with this newly invented instrument, was working in this invisible spectrum. His previous experience had been that of most scientific men—that very few discoveries come with a surprise and that they are usually the summation of the patient work of years.

In this case, almost the only one of his experience, he had the sensations of one who makes a discovery. He went down the spectrum, noting the evidence of invisible heat die out on the scale of the instrument until he came to the apparent end even of the invisible, beyond which the most prolonged researches of investigators up to that time had shown nothing. There he watched the indications grow fainter and fainter until they too ceased at the point where the French investigators believed they had found the very end of the end. By some happy thought he pushed the indications of this delicate instrument into the region still beyond. In the still air of this lofty region the sunbeams passed unimpeded by the mists of the lower earth, and the curve of heat, which had fallen to nothing, began to rise again. There was something there. For he found, suddenly and unexpectedly, a new spectrum of great extent, wholly unknown to science and whose presence was revealed by the new instrument, the bolometer.

This new spectrum is given on the map, where it will be observed that, while the work of the photograph (much more detailed than that of the bolometer, where it can be used at all) has been stated to extend, as far as regular mapping is concerned, to about $1 \cdot 1^{\mu}$, everything beyond this is due to the bolometer, except that early French investigators had found evidence of heat extending to $1 \cdot 8^{\mu}$. Still beyond that Ultima Thule, this region, which he has ventured to call the "New Spectrum", extends. It will be found between wave-lengths $1 \cdot 8^{\mu}$ and $5 \cdot 3^{\mu}$ on the map.

The speaker had been much indebted to others for the perfection to which the apparatus, and especially the galvanometer, had been brought. He was under obligations particularly to Mr. Abbot, for assistance in many ways, which he had tried to acknowledge in the volume; but before closing this most inadequate account of it, he would like to draw attention to one feature which was not represented in the spectrum map before them, although it would be found in the book.

During early years the impression had been made upon him that there were changes in the spectrum at different periods of the year. Some of these changes might be in the sun itself. The major portion of those he was immediately speaking of, he believed, were rather referable to absorptions in the earth's atmosphere.

Now these early impressions had been confirmed by the work of the Observatory in recent years, and charts given in the volume would show that (the sun being always supposed to be at about the same altitude, and its rays to traverse about the same absorbing quantity of the earth's atmosphere) the energy spectrum was distinctly different in spring, in summer, in autumn, and in winter. The lateness of the hour prevented him from enlarging on this latter profoundly interesting subject. He would only briefly point out the direction of these changes, which were not perhaps to be called conspicuous, but which seemed to be very clearly brought out as certainly existing. With regard to them he would only observe, what all would probably agree to, that while it has long been known that all life upon the earth, without exception, is maintained by the sun, it is only recently that we seem to be coming by various paths, and among them by steps such as these, to look forward to the possibility of a knowledge which has yet been hidden to us of the way in which the sun maintains it. We were hardly beginning to see yet how this could be done, but we were beginning to see that it might later be known, and to see that the seasons, which wrote their coming upon the records of the spectrum, might in the future have their effects upon the crops prevised by means somewhat similar to those previsions made day by day by the Weather Bureau, but in ways infinitely more far-reaching, and that these might be made from the direct study of the sun.

We are yet, it is true, far from able to prophesy as to coming years of plenty and of famine, but it is hardly too much to say that recent studies of others as well as of the writer strongly point in the direction of some such future power of prediction.

19

Spectroscopic Observations†

C. A. YOUNG

IN accordance with the programme, as soon as the sun came out into the clear sky, I had adjusted the slit of the spectroscope accurately tangential to the limb of the sun at the point where the last ray would vanish, and brought the 1474 line to the cross-wires. It was already plainly bright, the atmospheric glare being so much reduced as to make it perfectly easy to see.

The lines of *b* were also distinctly reversed, as were several of the iron lines near E; and I even thought that I could see the three chromium lines which I had found the day before.

Very soon, as the crescent grew narrower, they shone out unmistakably, and all the other lines I have mentioned became continually more conspicuous, while the dark lines of the spectrum and the spectrum itself gradually faded away; until all at once, as suddenly as a bursting rocket shoots out its stars, the whole field of view was filled with bright lines more numerous than one could count.

The phenomenon was so sudden, so unexpected,‡ and so wonderfully beautiful, as to force an involuntary exclamation.

---

† *Coast Survey Report*, Appendix 16, 1870. Reprinted from *Mem. Roy. Astron. Soc.* **41**, 434, 1879.

‡ It was unexpected simply because it had not been seen in 1868 and 1869. In 1869, having been led to expect something of the kind by Father Secchi's report of a layer close to the sun's surface, giving a continuous spectrum, I looked for it very carefully, but failed to see it; so that on this occasion I was wholly unprepared.

I now suppose that my previous failure was due to my having worked with a *radial* slit; in this case the lines would be so short (from 0·5″ to 1·5″) that they might easily escape observation.

Gently and yet very rapidly they faded away, until within about two seconds, as nearly as I can estimate, they had vanished, and there remained only the few lines I had observed at first.

Of course it would be very rash, on the strength of such a glimpse, to assert with positiveness that these innumerable lines corresponded exactly with the dark lines of the spectrum which they replaced; but I feel pretty fairly confident that such was the case.

The grouping of the lines seemed perfectly familiar, and so did the general appearance of the spectrum; except that the lines which had been visible before the totality were relatively far too conspicuous.

20

Observations on the
Solar Eclipse of 12 Dec. 1871†

P. J. C. JANSSEN

Sholoor, December 19, 1871.

I HAD the honour of sending you on the very day of the eclipse
a few lines to inform the Academy that I had observed the eclipse
under an exceptional sky, and that my observations led me to
assume a solar origin for the Corona (see *Nature*, **5,** 190).
Immediately after the eclipse I was obliged to busy myself with
the personal and material arrangements for my expedition into
the mountains, and hence I have been unable to complete any
detailed account, but I take advantage of the departure of this
courier to give some indispensable details as to the announced
results. Without entering into a discussion, which will form part
of my narrative, I shall say, in the first place, that the magnificent
Corona observed at Sholoor showed itself under such an aspect
that it seemed to me impossible to accept for it any cause of the
nature of the phenomena of diffraction or reflection upon the
globe of the moon, or of simple illumination of the terrestrial
atmosphere. But the arguments which militate in favour of an
objective and circumsolar cause, acquire invincible force when
we inquire into the luminous elements of the phenomenon. In
fact, the spectrum of the Corona appeared in my telescope, not
continuous, as it had previously been found, but remarkably
complex. I detected in it, though much weaker, the brilliant lines
of hydrogen gas, which forms the principal element of the protuberances and chromosphere; the brilliant line which has already
been indicated during the eclipses of 1869 and 1870, and some

† *Nature* **5,** 249, 1872.

other fainter ones; obscure lines of the ordinary solar spectrum, especially that of sodium (D); these lines are much more difficult to perceive. These facts prove the existence of matter in the vicinity of the sun; matter which manifests itself in total eclipses by phenomena of emission, absorption, and polarisation. But the discussion of the facts leads us still further. Besides the cosmical matter independent of the sun which must exist in its neighbourhood, the observations demonstrate the existence of an excessively rare atmosphere, with a base of hydrogen, extending far beyond the chromosphere and protuberances, and deriving its supplies from the very matter of the latter—matter which is projected with so much violence, as we may ascertain every day. The rarity of this atmosphere at a certain distance from the chromosphere must be excessive; its existence, therefore, is not in disagreement with the observations of some passages of comets close to the sun.

21

The Wave-length of the Green Coronal Line and Other Data Resulting from an Attempt to Determine the Law of Rotation of the Solar Corona†

W. W. CAMPBELL

A DETERMINATION of the law of rotation of the solar corona would no doubt be valuable on many accounts. Interest in this problem was aroused by Deslandres' attempt to solve it at the 1893 eclipse, in Senegal. The reality of his result has been questioned on the ground that the H and K calcium lines, used by him, do not have their origin in the corona, but in the prominences and chromosphere. It seemed proper that another determination should be attempted, basing it upon light radiations which are of unquestioned coronal origin. Accordingly, as one of many problems, it was undertaken by the Crocker expedition‡ sent out from the Lick Observatory to observe the eclipse of January 22, 1898, in India.

It was evident that this investigation, involving the Doppler-Fizeau principle, and requiring high dispersion, could apply, with any hope of success, only to the bright-line portion of the corona. Existing data seemed to show that the green line and those near $\lambda\lambda$ 423 and 399 were the only reasonably strong and unquestioned coronal lines in the available portion of the spectrum; and further, that the photographic action of the green radiation was vastly

† *Astrophys. Journ.* **10**, 186–92, 1899. Read at the Third Conference of Astronomers and Astrophysicists, September 7, 1899.

‡ The expenses of the expedition were defrayed by the late Hon. C. F. Crocker, a regent of the University of California. He was also the patron of the expeditions to Cayenne (1889) and Japan (1896).

stronger than in the case of the other two lines, even though the green line lay in a region of weakness on isochromatic plates. It was therefore decided to base the observations on the green line. The justification of this decision seems clear, in view of Newall's experience.† His ingenious and powerful apparatus was constructed for recording the blue line at λ423; but this plate, on development, showed no trace of any impress of the coronal line.

Since prismatic dispersion in the green is relatively weak, the question of using a large-sized grating was considered. Two gratings with ruled surfaces about 2·5 × 3·9 inches were kindly loaned by Mr. Brashear; but some simple experiments indicated clearly that a large number of suitable prisms would be more efficient for this purpose than a grating, besides permitting greater compactness and stability of mounting.

The optical train of the instrument employed was as follows:

Image lens; clear aperture, $2\frac{1}{16}$ inches; focal length, $19\frac{7}{16}$ inches; forming an image of the Sun on the slit, with diameter 0·184 inch.

Collimator lens; clear aperture, $2\frac{5}{32}$ inches; focal length, $20\frac{3}{4}$ inches.

Four compound prisms, altitude of each, $2\frac{5}{8}$ inches; face, $2\frac{1}{8}$ inches; combined minimum deviation for λ5317, 152°37′.

Two single 60° prisms; altitude, $2\frac{1}{2}$ inches; face, $3\frac{1}{8}$ inches; combined minimum deviation for λ5317, 113°04′.

Camera lens; clear aperture, $2\frac{1}{2}$ inches; focal length, 20 inches.

The combined deviation of the six prisms being 265°41′, the beam of light in the camera crossed the beam in the collimator nearly at right angles.

These optical parts were mounted in wood (Spanish cedar) from my drawings, by the Observatory carpenter, and finished in shellac. The instrument was easily adjusted, and worked well.

Immediately in front of the plate-holder were three rectangular sliding diaphragms for controlling the exposures. These could be withdrawn, singly, in a direction parallel to the Fraunhofer lines of the spectrum. One of these (A) covered the red end of the

† *Proc. Royal Soc.* **64**, 56, 1898.

spectrum up to within $\frac{1}{4}$ inch of $\lambda5317$. Another (B) covered the violet end up to within $\frac{1}{4}$ inch of $\lambda5317$. A third (C), about $\frac{5}{8}$ inch wide, covered the central half inch of the plate, overlapping (A) and (B) very slightly.

A Cramer isochromatic plate, backed with Carbutt's liquid backing, was employed.

The slit of the spectrograph was set at 0·04 mm and directed upon the solar equator. About 15 minutes before totality the red end of the plate was uncovered by withdrawing diaphragm A. The solar crescent was caused to drift rapidly along the slit, thereby recording the solar spectrum in the region A, for reference. Diaphragm A was inserted, and the polar axis was set in motion by the clockwork. Two seconds after totality, diaphragm C was withdrawn, allowing the region of the coronal spectrum between $\lambda\lambda5260$ and 5430 to record itself on the plate. The diaphragm was inserted at 1^m 52^s after the beginning of totality. After third contact, diaphragm B was withdrawn, and the solar crescent caused to drift along the slit, recording the solar spectrum on the violet end of the plate, for reference.

On developing the plate, in camp, the solar spectrum at the red end of the plate was found to be suitably exposed, whereas that at the violet end was underexposed—but measurable—probably on account of my underestimating the reduced brightness of the thin crescent. The continuous spectrum of the inner corona was strongly recorded, as was also one strong bright line. However, I was struck with the fact that the bright line fell much further to the violet side of the region uncovered by diaphram C than I had expected. Being busy with further photographic developing, in the intense heat, the matter escaped my attention. While engaged in measuring this plate, in January 1899, I learned that Lockyer had assigned a new wave-length to the green coronal line. Reducing my measures, I at once obtained a result in substantial agreement with his.

The details of my determination of the wave-length are given below (Table 1). The first column contains Rowland's wave-lengths of the solar lines used for reference. The second contains these wave-lengths corrected for the fact that the lines under diaphragm A had their origin at the E. (approaching) limb of the Sun and those under B had their origin at the W. (receding) limb. The micrometer readings on the solar lines, and on the bright lines, are contained in the third and fourth columns, the value of a revolution of the screw being 0·25 mm. The readings on the bright coronal line were made at points 2′ from the E. limb and 1′ from the W. limb.

TABLE 1

Rowland's W.-l. in ⊙	Corrected W.-l. in ⊙	Micrometer measures E. of ⊙	Micrometer measures W. of ⊙	Computed W.-l.	
				E. of ⊙	W. of ⊙
5183·79	5183·83	12·162	12·062	5183·83	5183·83
5227·2⁴	5227·2†	27·883	27·790	5227·26	5227·21
Bright Line		53·176	53·185	5303·21	5303·32
5463·33	5463·29	98·908	98·957	5463·30	5463·36
5476·9⁴	5476·9⁴	102·402	102·427	5476·96	5476·90
5497·74	5497·70	107·603	107·650	5497·70	5497·70
5528·64	5528·60	115·137	115·177	5528·67	5528·60
5603·10	5603·06	132·161	132·226	5603·02	5602·95
5688·44	5688·40	150·080	150·152	5688·53	5688·38
5763·22	5763·18	164·441	164·562	5763·18	5763·18

† A double line.

The wave-length of the bright line was computed by means of Dr. Hartmann's exceedingly valuable interpolation formula,†

$$\lambda - \lambda_0 - \frac{C}{R_0 - R} = 0.$$

Substituting 5183·83, 5497·70 and 5763·18 successively for λ, and their corresponding micrometer readings for R, and solving for λ_0, C and R_0, there resulted,

$$\text{For E. side, } \lambda = 3805 \cdot 43 + \frac{[5 \cdot 850829]}{526 \cdot 744 - R};$$

$$\text{For W. side, } \lambda = 3806 \cdot 57 + \frac{[5 \cdot 850844]}{527 \cdot 088 - R}.$$

† This journal **8**, 218, Nov. 1898.

Substituting the various values of R in these equations, and solving for λ, we readily obtain the wave-lengths in the last two columns of the table. The three wave-lengths on which the formulae were based naturally reproduce themselves. The accordance of the computed values of the wave-lengths with Rowland's values furnishes an indication of the accuracy of the results. No doubt a slight improvement of the residuals would have resulted from a least square determination of the values of λ, C and R_0; but on account of the character of the bright line, to be described later, this would have been superfluous.

The wave-lengths obtained for the coronal line are

For E. side, 5303·21
For W. side, 5303·32
Mean, 5303·26

The difference of the determinations for the E. and W. sides is 0·11 t.m., corresponding to a relative velocity in the line of sight, for the two sides, of 6·2 km, or a rotational velocity of 3·1 km per second. However, I regard this last result as subject to a possible error of at least ± 2 km per second: partly on account of unavoidable errors of observation, but principally on account of the character of the bright line. The inner ends of the bright line are overexposed, while its outer ends are underexposed, and it does not seem to be monochromatic. The suitably exposed portions of the line are not only ill-defined, but unsymmetrical, and accurate settings on them could not be made. It is possible that the original radiations were reasonably monochromatic. A good photograph of the green coronal ring was secured with another of our instruments, an objective grating spectrograph, showing this ring to be extremely irregular in form; and I believe the observations by Lockyer, Newall and others led to the same conclusion. Such being the case, we should expect rapid movements to occur within this atmosphere. The recorded appearance of the line $\lambda5303$ finds, possibly, an easy explanation in the distortions due to relative velocities, in the line of sight, of the different portions projected on the slit.

In planning the apparatus, it was taken for granted that this line would fall at λ5317. Inasmuch as some of the earlier observers mention having seen one or two additional bright lines in this vicinity, the diaphragms were arranged so that one half inch of the plate was reserved for coronal exposure, hoping to record any of these additional lines. This is responsible for the great distance between the bright line and the solar reference lines; and, besides, the whole purpose of the main problem was to measure *difference* of wave-lengths. Had it been intended to determine the wave-length of the green line, the comparison spectrum would have been arranged very differently. However, the above value of the wave-length should not be in error by more than ±0·15 t.m.

As confirmatory of the above value, the green ring on the objective grating spectrograph, referred to above and to be described later, is at λ5303.

It is desirable that the position of the line should be determined as accurately as possible at the next eclipse.

The radial length of recorded bright line at the E. limb is 4', and at the W. limb 2'. It does not seem wise to attempt to measure the coronal rotation at the short eclipse of 1900, but it is possible that useful results might be obtained at the East Indian eclipse of 1901, lasting 6½ minutes.

The continuous spectrum of the inner corona was recorded out to a distance of 2·5' on the E. side, and of 1·5' on the W. side. The greater strength of coronal radiation on the E. side is very apparent for both bright-line and continuous spectra. While the dark lines in the recorded comparison spectra are sharp and strong, there is not the slightest trace of dark lines in the recorded continuous spectrum of the corona. This radiation seems to be of coronal origin, and not due to reflected photospheric radiations.

I do not think it is difficult to explain the origin of the error which has prevailed for many years in the accepted value (λ5316·87) of the wave-length of the bright coronal line; and the following explanation is respectfully suggested.

The strongest chromosphere line in this region is the one at λ5317. On one of my photographs, giving a continuous record of the spectrum of the Sun's edge, both when the thin crescent was gradually disappearing at contact II, and reappearing at contact III, the line λ5317 is the brightest in this region, and "persisted" slightly longest.† Likewise, in visual observations at contact II, it would no doubt be the brightest line visible, and "persist" longest. Inasmuch as my moving plate recorded many of the faintest chromosphere lines in Professor Young's list, but made no record of a bright line at λ5303, it is pretty certain that the true coronal line would be difficult to observe so long as the chromosphere spectrum was visible. The observers made it their first duty to fix the position of the green line. The persisting chromosphere line, very conspicuous just before and at the instant of totality, was naturally assumed to be identical with the true coronal line, and its position was fixed at 1474K. Later, when this line had disappeared, rather suddenly, and the background had become dark enough to allow the line at λ5303 to be seen, the observers were interested in determining the extent, and other allied properties, of this line; and no further micrometer settings were made for determining its wave-length. This illustrates one of the advantages of photographic methods, now happily available.

The photograph above described is not suitable for mechanical reproduction; but, with the Director's assent, I should be willing to supply copies on glass to those investigators who are planning for observations, based on the green coronal line.

Professor Young contributed largely to this determination of the position of the coronal line, by arranging for the most generous loan of the four large compound prisms, described above, and belonging to Princeton University; and also, by the fact that the instrument was manipulated during totality, in a perfectly satisfactory manner, by one of his former students in Dartmouth College, the Rev. J. E. Abbott, long a resident of Bombay, who extended many favors to the expedition.

† This persistence may be due to the greater brightness of the line, rather than to a thicker stratum.

22

Spectroscopic Observations of the Sun IV†

J. Norman Lockyer

I beg to lay before the Royal Society very briefly the results of observations made on the 11th instant in the neighbourhood of a fine spot, situated not very far from the sun's limb.

   I. Under certain conditions the C and F lines may be observed *bright on the sun*, and in the spot-spectrum also, as in prominences or in the chromosphere.

  II. Under certain conditions, although they are not observed as bright lines, the corresponding Fraunhofer lines are blotted out.

 III. The accompanying changes of refrangibility of the lines in question show that the absorbing material moves upwards and downwards as regards the radiating material, and that these motions may be determined with considerable accuracy.

 IV. The bright lines observable in the ordinary spectrum are sometimes interrupted by the spot-spectrum. *i.e.* they are only visible in those parts of the solar spectrum near, and away from, spots.

  V. The C and F lines vary excessively in thickness over and near a spot, and on the 11th in the deeper portion of the spot they were much thicker than usual.

 VI. Stars, in the spectrum of which the absorption-lines of hydrogen are absent, may either have their chromospheric light radiated from beyond the limb just balanced by the

† *Proc. Roy. Soc.* **17**, 415–18, 1869.

light absorbed by the chromosphere on the disk, or they may come under the condition referred to in (II.), either absolutely or on the average.

Addendum

Since the date on which the foregoing paper was written, I have obtained additional evidence on the points referred to. I beg therefore to be permitted to make the following additions to it.

The possibility of our being able to determine the velocity of movements of uprush and downrush taking place in the chromosphere depends upon the alterations of wave-length observed.

It is clear therefore that a mere uprush or downrush at the sun's limb will not affect the wave-length, but that if we have at the limb cyclones, or backward or forward movements, the wave-length will be altered; so that we may have:—

1. An alteration of wave-length near the centre of the disk caused by upward or downward movements.
2. An alteration of wave-length close to the limb, caused by backward or forward movements.

If the hydrogen-lines were invariably observed to broaden out on both sides, the idea of movement would require to be received with great caution; we might be in presence of phenomena due to greater pressure, both when the lines observed are bright or black upon the sun; but when they widen out sometimes on one side, sometimes on the other, and sometimes on both, this explanation appears to be untenable, as Dr. Frankland and myself in our researches at the College of Chemistry have never failed to observe a widening out on both sides the F line when the pressure of the gas has been increased.

On the 21st I was enabled to extend my former observations.

On that day the spot, observations of which form the subject of the paper, was very near the limb; as this was the first opportunity of observing a fine spot under such circumstances I had been able to utilize, I at once commenced work upon it. The spot was so near the limb that its spectrum and that of the chromosphere were both visible in the field of view.

The spot-spectrum was very narrow, as the spot itself was so greatly foreshortened; but the spectrum of the chromosphere showed me that the whole adjacent limb was covered with prominences of various heights all blended together.

Further, the prominences seemed fed, so to speak, from, apparently, the preceding edge of the spot; for both C, F, and the line near D, *were magnificently bright on the sun itself*, the latter especially striking me with its thickness and brilliancy.

In the prominences C and F were observed to be strangely gnarled, knotty, and irregular, and I thought at once that some "injection" must be taking place. I was not mistaken. On turning to the magnesium lines I saw them far above the spectrum of the limb and unconnected with it.

A portion of the upper layer of the photosphere had in fact been lifted up beyond the usual limits of the chromosphere, and was there floating cloud-like.

The vapour of sodium was also present in the chromosphere, though not so high as the magnesium, or unconnected with the spectrum of the limb, and, as I expected, with such a tremendous uplifting force, I saw the iron lines (for the first time) in the spectrum of the chromosphere.

My observations commenced at 7.30 a.m.; by 8.30 there was comparative quiet.

At 9.30 the action had commenced afresh; there was not a single prominence.

At the base of the prominence I got this appearance:

Higher up this:　　　. Here I may be permitted to recall the observation made of March 14, in which a slight movement of the slit gave me first 　　　, then 　　　, and finally 　　　,

all these appearances being due to cyclonic action.

On the following side of the spot, at about 10 a.m., I observed that the F line had disappeared; at the point of disappearance there appeared to be an elongated brilliantly illuminated lozenge lying across it at right angles, as if the spectroscope were analyzing the light proceeding from a cyclone of hydrogen on the sun itself, but so near the limb that the rotatory motion could be detected.

The next observations I have to lay before the Royal Society were made on the 27th inst. Careful observations on the 25th and 26th revealed nothing remarkable except that the chromosphere was unusually uniform.

On the 27th a fine spot with a long train of smaller ones and faculae was well on the disk. The photosphere in advance of the spot, and the large spot itself, showed no alteration from the usual appearance of the hydrogen-lines; but in the tails of the spot the case was widely different.

The F line, at which I worked generally, as the changes of wavelength are better seen, was as irregular as on the former occasions.

    I. It often stopped short of one of the small spots, swelling out prior to disappearance.

    II. It was invisible in a facula between two small spots.

    III. *It was changed into a bright line, and widened out on both sides two or three times* IN THE VERY SMALL SPOTS.

    IV. Once I observed it to become bright *near* a spot, and to expand over it on both sides.

    V. Very many times near a spot it widened out, sometimes considerably, on the less refrangible side.

    VI. Once it extended as a bright line without any thickening over a small spot.

    VII. Once it put on this appearance: bright.

    VIII. I observed in it all gradations of darkness.

    IX. When the bright and dark lines were alongside, the latter was always the less refrangible.

23

The Spectroheliograph†

G. E. HALE

THE spectroscopic method, as applied by astrophysicists in various parts of the world, has yielded a nearly continuous record of the solar prominences extending back over more than thirty years. For many purposes such a record is entirely satisfactory, and permits important conclusions to be drawn. But the process of observation is not only slow and painstaking: it is subject to the errors and uncertainties that necessarily attend the hand delineation of any object, seen through a fluctuating atmosphere. Moreover, changes in the forms of eruptive prominences are frequently so rapid that the draughtsman cannot record them. It was principally in the hope of simplifying the process of observation, and of rendering it more rapid and more accurate, that the spectroheliograph was devised at the Kenwood Observatory in 1889.‡

The principle of this instrument is very simple. Its object is to build up on a photographic plate a picture of the solar flames, by recording side by side images of the bright spectral lines which characterize the luminous gases. In the first place, an image of the Sun is formed by a telescope on the slit of a spectroscope. The light of the Sun, after transmission through the spectroscope, is spread out into a long band of color, crossed by lines representing

† *Stellar Evolution*, chapter XI, pp. 82–96, University of Chicago Press, 1908.

‡ It was subsequently learned that the method embodied in the spectroheliograph had been suggested by Janssen as early as 1869, reinvented by Braun of Kalocsa, and actually tried by Lohse at Potsdam. But it had not proved a success.

the various elements. At points where the slit of the spectroscope happens to intersect a gaseous prominence, the bright lines of hydrogen and helium may be seen extending from the base of the prominence to its outer boundary. If a series of such lines, corresponding to different positions of the slit on the image of the prominence, were registered side by side on a photographic plate, it is obvious that they would give a representation of the form of the prominence itself. To accomplish this result, it is necessary to cause the solar image to move at a uniform rate across the first slit of the spectroscope, and, with the aid of a second slit (which occupies the place of the ordinary eyepiece of the spectroscope), to isolate one of the lines, permitting the light from this line, and from no other portion of the spectrum, to pass through the second slit to a photographic plate. If the plate be moved at the same speed with which the solar image passes across the first slit, an image of the prominence will be recorded upon it. The principle of the instrument thus lies in photographing the prominence through a narrow slit, from which all light is excluded except that which is characteristic of the prominence itself. It is evidently immaterial whether the solar image and photographic plate are moved with respect to the spectroheliograph slits, or the slits with respect to a fixed solar image and plate.

This method, when first tried at the Harvard Observatory in 1890, proved unsuccessful. The lack of success was partly due to the fact that a line of hydrogen was employed. This line, though fairly suitable for the photography of prominences with the perfected spectroheliograph of the present day, was too faint for successful use amidst the difficulties which surrounded the first experiments. Accordingly, when the work was resumed a year later at the Kenwood Observatory in Chicago an attempt was made, through a photographic investigation of the violet and ultra-violet regions of the prominence spectrum, to discover other lines better fitted for future experiments. In the extreme violet region, in the midst of two broad dark bands which form the most striking feature of the solar spectrum, two bright lines (H and K) were found and attributed to the vapor of calcium. They had previously been seen visually by Young, but, on account of the insensitiveness of the eye for light of this color, they could not be observed satisfactorily. A careful study soon showed them to be present in every prominence examined, at elevations above the solar surface equaling or exceeding those attained by hydrogen itself. Their suitability for the purpose of prominence photography is due to several causes, among which may be mentioned their exceptional brilliancy, their presence at the center of broad dark bands which greatly diminish the brightness of the sky spectrum, and the comparatively high sensitiveness of photographic plates for light of this wave-length.

While fairly efficient from an optical point of view, the spectroheliograph of the preceding year had possessed many mechanical defects. It sufficed to give photographs of individual prominences, but they were not very satisfactory. In a new instrument, devised for use with the 12-inch Kenwood telescope, the principal defects were overcome, and means of securing the necessary conditions of the experiment were provided. In this instrument the solar image and photographic plate were fixed, while the first and second slits were made to move across them by means of a system of levers, set in motion by hydraulic power. The first trials of the instrument, made in January, 1892, were entirely successful, and the chromosphere and prominences surrounding the Sun's disk were easily and rapidly recorded. The details of their structure were shown with the sharpness and precision characteristic of the best eclipse photographs. And the opportunity for making such records, previously limited to the brief duration, never exceeding seven minutes, of a total eclipse, was at once indefinitely extended. Thus it became possible to study photographically the slowly varying forms of the quiescent, cloudlike prominences, and, to particular advantage, the rapid changes of violent eruptions.

But even before this primary purpose of the work had been accomplished, the possibility of making another and much more important application of the instrument had presented itself. A photographic study of the spectrum of various portions of the Sun's surface had shown the existence at many points of great clouds of calcium vapor, luminous enough to render their existence evident through the production of bright H and K lines on the solar disk. Some of these calcium clouds had, indeed, been known to exist through the important visual observations of Young, who had observed the bright H and K lines in the vicinity of sun-spots. But the vast extent and the characteristic forms of the phenomena could not be ascertained by such means. What was required was such a representation of the solar disk as the spectroheliograph had been designed to give in the case of the prominences. From a consideration of the results obtained in the spectroscopic study of the disk, it appeared probable that an important application of the spectroheliograph might be made in this new direction.

Before describing this second application of the instrument, it may be well to recall the appearance of the Sun when seen with a telescope, or when photographed in the ordinary manner without a spectroheliograph. From photographs like these, we see that the most conspicuous features of the solar surface, at least so far as the eye can detect, are the well-known sun-spots. The bright faculae, which rise above the photospheres, are conspicuous when near the edge of the Sun, but practically invisible when they happen to lie near the center of the disk. The bright H and K lines, referred to in the last paragraph, were found in close association with the faculae, and it appeared probable that much of the highly heated calcium vapor, to which these bright lines are due, rises from the interior of the Sun through the faculae. It was therefore to be expected that a successful application of the spectroheliograph to the photography of the luminous calcium clouds would give bright forms resembling those of the faculae. Furthermore, it was to be hoped that these brilliant clouds could be recorded, not only near the limb of the Sun, but also in the central part of the disk, since the bright reversals of the H and K lines were equally well photographed in all parts of the image.

The results of the first experiments, which were made at the beginning of 1892, were such as to justify fully the expectations that had been entertained. It was at once found possible to record the forms, not only of the brilliant clouds of calcium vapor associated with the faculae, and occurring in the vicinity of sun-spots, but also of a reticulated structure extending over the entire surface of the Sun. The earliest use of the method was made in the study of the great sun-spot of February 1892, which, through the great scale of the phenomena it exhibited and the rapid changes that resulted from its exceptional activity, afforded the very conditions required to bring out the peculiar advantages of the spectroheliograph. In the systematic use of the instrument continued at the Kenwood Observatory through the following years, a great variety of solar phenomena were recorded, and the changes which they underwent from day to day—sometimes, in the more violent eruptions, from minute to minute—were registered in permanent form. During this period, which ended with the transfer of the Kenwood instruments to the Yerkes Observatory, over 3000 photographs of solar phenomena were secured. From a systematic study of these negatives, in the course of which the heliographic latitude and longitude of the calcium clouds (subsequently named the *flocculi*) in many parts of the Sun's disk were measured from day to day (by Fox), a new determination of the rate of the solar rotation in various latitudes has been made. This shows that the calcium flocculi, like the sun-spots, complete a rotation in much shorter time at the solar equator than at points nearer the poles. In other words, the Sun does not rotate as a solid body would do, but rather like a ball of vapor, subject to laws which are not yet understood.

In this first period of its career the spectroheliograph had therefore permitted the accomplishment of two principal objects. It had provided a simple and accurate means of photographing the solar prominences in full sunlight, which gave results hardly inferior to those obtained during the brief moments of a total eclipse. It had also given a means of recording a new class of phenomena, known previously to exist only through glimpses of the bright calcium lines in the vicinity of sun-spots, but wholly invisible to observation, either visually or on photographs taken by ordinary methods. It was not difficult to see, however, that the possibilities of the new method were much greater than had been indicated by the work so far accomplished. It seemed probable that our knowledge of the finer details of the calcium flocculi would be greatly increased if provision could be made for photographing a much larger solar image with a spectroheliograph of improved design. And it was furthermore evident that other applications of the instrument, involving the use of different spectral lines, and the employment of principles which had not been thoroughly tested in the earlier work, might reasonably be hoped for. Attempts were, indeed, made to photograph the Sun's disk with the dark lines of hydrogen, but the Kenwood spectroheliograph was not well adapted for this purpose.

The 40-inch telescope of the Yerkes Observatory provided the first requisite for the new work—namely, a large solar image, having a diameter of 7 inches as compared with the 2-inch image given by the Kenwood telescope. The construction of a spectroheliograph large enough to photograph such an image of the Sun involved serious difficulties, but these were finally overcome. The Rumford spectroheliograph, designed to meet the special conditions of the new work, was built in the instrument shop of the Yerkes Observatory, and is now in daily use with the 40-inch telescope.

In this instrument the solar image is caused to move across the first slit by means of an electric motor, which gives the entire telescope a slow and uniform motion in declination. The sunlight, after passing through the first slit, is rendered parallel by a large lens at the lower end of the collimator tube. The parallel rays from this lens fall upon a silvered glass mirror, from which they are reflected to the first of two prisms, by which they are dispersed into a spectrum. After passing through the prisms, the light, which has now been deflected through an angle of 180°, falls upon a second large lens at the lower end of the camera tube. This forms an image of the spectrum at the upper end of the tube, where the second slit is placed. Any line in the spectrum may be made to fall upon this slit, by properly adjusting the mirror and prisms. Above the slit, and nearly in contact with it, the photographic plate is mounted in a carriage, which runs on rails at right angles to the length of the slit. The rails are covered by a light-tight camera box, so that no light can reach the plate except that which passes through the second slit. While the solar image is moving across the first slit, the plate is moved at the same rate across the second slit, by a shaft leading down the tube from the electric motor, and connected, by means of belting, with screws that drive the plate-carriage.

Photographs of the solar disk taken with this instrument under good atmospheric conditions reveal a multiplicity of fine details. The entire surface of the Sun is shown by these plates to be covered by minute luminous clouds of calcium vapor, only about a second of arc in diameter, separated by darker spaces, and closely resembling in appearance the well-known granulation of the solar photosphere. A sharp distinction must, however, be drawn between this appearance, which is wholly invisible to the eye at the telescope, and the granulation of the photosphere. In accordance with Langley's view, the grains into which the solar surface is resolved under good conditions of visual observation are the extremities of columns of vapor rising from the Sun's interior. They seem to mark the regions at which convection currents, proceeding from within the Sun, bring up highly heated vapors to a height where the temperature becomes low enough to permit them to condense. It might be anticipated that out of the summits of these condensed columns, other vapors, less easily condensed, would continue to rise, and that the granulated appearance obtained with the spectroheliograph may represent the calcium clouds thus ascending from the columns. We might, indeed, go a step farther, and imagine the larger and higher calcium clouds to be constituted of similar vaporous columns, passing upward through the chromosphere, and perhaps at times extending out into the prominences themselves. A means of research now to be described, which represents another application of the spectroheliograph, involving a new principle, seems competent to throw some light on this question.

Mention has already been made of the faculae, which are simply regions in the photosphere that rise above the ordinary level. Near the edge of the Sun their summits lie above the lower and denser part of that absorbing atmosphere which so greatly reduces the Sun's light near the limb, and in this region the faculae may be seen visually. At times they may be traced to considerable distances from the limb, but as a rule they are inconspicuous or wholly invisible toward the central part of the solar disk. The Kenwood experiments had shown that the calcium vapor coincides closely in form and position with the faculae, and hence the calcium clouds were long spoken of under this name. In the new work at the Yerkes Observatory the differences between the calcium clouds and the underlying faculae became so marked that a distinctive name for the vaporous clouds appeared necessary. They were therefore designated *flocculi*, a name chosen without reference to their particular nature, but suggested by the flocculent appearance of the photographs.

In order to analyze these flocculi and to determine their true structure, a method was desired which would permit sections of them at different heights above the photosphere to be photographed. Fortunately there is a simple means (first described by Deslandres) which appears to accomplish this apparently difficult object. At the base of the flocculi the calcium vapor, just rising from the Sun's interior, is comparatively dense. As it passes upward through the flocculi it reaches a region of much lower pressure, and during the ascent it might be expected to expand, and therefore to become less dense. Now we know from experiments in the laboratory that dense calcium vapor produces very broad spectral bands, and that, as the density of the vapor is decreased, these bands narrow down into fine sharp lines. An examination of the solar spectrum will show that the H and K lines of calcium give evidence of the occurrence of this substance under widely different densities in the Sun. The broad dark bands, which for convenience we designate H_1 and K_1, are due to the low-lying, dense calcium vapor. At their middle points (over flocculi) are seen two bright lines, which are much narrower and better defined. These lines, designated H_2 and K_2, are the ones ordinarily employed in photographing the flocculi with the spectroheliograph. Superposed upon these bright lines are still narrower dark lines, due to the absorption of cooler calcium vapor at higher elevations (H_3, K_3). It will be seen that the evidence of the existence of calcium vapor at various densities in the Sun is apparently complete, and that we may here find a way of photographing the vapor at low levels without admitting to the photographic plate any light that comes from the rarer vapors at higher levels. It is simply necessary to set the second slit of the spectroheliograph near the edge of the board H_1 or K_1 bands, in order to obtain a picture showing only that vapor which is dense enough to produce a band of width sufficient to reach this position of the slit. No light from the rarer vapors above can enter the second slit under these circumstances, since they are incapable of producing a band of the necessary width.†

The great sun-spot of October, 1903, afforded an opportunity to try this method in a very satisfactory manner. The manner in which the vapor at the H_2 level overhangs the edge of the sun-spot is very striking, and thorough study should throw some light on the conditions which exist in such regions. For it is possible, not only to photograph sections of the vapor at various levels, but also to ascertain, by the displacement of the H_2 or H_3 line, as photographed by a powerful spectrograph, the direction and velocity of motion of the vapor which constitutes the flocculi. It is commonly found that the vapor is moving upward at the rate of about one kilometer per second, though the velocity varies considerably at different points and under different conditions.

The photographs occasionally show the existence of flocculi remarkable for their great brilliancy. In these regions active eruptions are in progress. The vapor, rendered highly luminous

† The bright regions photographed in this way resemble the faculae very closely, and may be regarded as essentially identical with them, since the white light from the continuous spectrum of the faculae contributes in an important degree to the formation of the photographic images. However, any dense calcium vapor which extends beyond the boundaries of the faculae will be recorded on the photograph. In any case we should expect the dense calcium vapor, supposed to be rising from the faculae, to correspond closely with them in form.

by intense heat or other causes, is shot out from the Sun's interior with great velocity. Consequently there are rapid changes in the forms of these brilliant regions, whereas the ordinary flocculi change slowly, and represent a much less highly disturbed condition of affairs. The brilliant eruptive flocculi always occur in active regions of the solar surface, and probably correspond with the eruptive prominences sometimes photographed projecting from the Sun's limb. A remarkable instance was recorded on the Kenwood photographs, which showed four successive stages of an eruption of calcium vapor on an enormous scale. A vast cloud thrown out from the Sun's interior completely blotted from view a large sun-spot, and spread out in a few minutes so as to cover an area of four hundred millions of square miles.

Although the eruptive flocculi probably correspond in many instances with eruptive prominences, it must not be concluded that the quiescent calcium flocculi correspond with the quiescent, cloudlike prominences. As a matter of fact, we have good evidence for the belief that the flocculi shown in these photographs represent in most instances comparatively low-lying vapors, while the prominences, which extend above the level of the chromosphere, do not ordinarily reveal themselves as bright objects in projection against the disk.

So far, we have considered the photography of the Sun with the light of the H and K lines of calcium. But it must naturally occur to anyone familiar with the solar spectrum that it should be possible to take photographs corresponding to other lines, and thus representing the vapors of other substances. For the darkness of the lines is only relative; if they could be seen apart from the bright background of continuous spectrum on which they lie, these lines would shine with great brilliancy. It is thus evident that, if all light except that which comes from one of these dark lines can be excluded from the photographic plate by means of the second slit of the spectroheliograph, it should be possible to obtain a photograph showing the distribution of the vapors corresponding to the line in question.

At this point attention should be called to the extreme sensitiveness of the spectroheliograph in recording minute variations in the intensity of a line—variations so slight that no trace of them can be seen in a spectrum photograph showing only the line itself. A well-known physiological effect is here concerned, for it is common experience that the eye cannot detect minute differences of intensity in various parts of an extremely narrow line, whereas these would become conspicuous if the line were widened out into a band of considerable width. The spectroheliograph records side by side upon the photographic plate a great number of images of a line which, taken together, build up the form of the region from which the light proceeds. In this way the full benefit of the physiological principle is derived, and very minute differences of intensity at various parts of the solar disk are clearly registered upon the plate.

It is obviously essential in photographing with the dark lines to exclude completely the light from the continuous spectrum on either side of the line employed. The admission of even a small quantity of this light might completely nullify the slight differences of intensity recorded by the aid of the comparatively faint light of the dark line. As the second slit cannot be narrowed beyond a certain point, it is evident that for successful photography with the dark lines their width must be increased by dispersion in the spectroheliograph to such a degree as to make them wider than the second slit.

The first satisfactory photographs obtained with dark lines were made with the Rumford spectroheliograph in May, 1903. The lines of hydrogen were chosen for this purpose, on account of their considerable breadth, and because of the prominent part played by this gas in the chromosphere and prominences. In order to secure sufficient width of the lines, the mirror of the spectroheliograph was replaced by a large plane grating having 20,000 lines to the inch. After leaving the grating the diffracted light enters the prisms, where it is still further dispersed before the image of the spectrum is formed upon the second slit. The effect of the prisms is not only to give additional dispersion, but also to reduce the intensity of the diffuse light from the grating—a most important matter in work of this nature. The hydrogen lines employed were $H\beta$, $H\gamma$, or $H\delta$, in the green-blue, blue, and violet, respectively.

On developing the first plate it was surprising to find evidences of a mottled structure covering the Sun's disk, resembling in a general way the structure of the calcium flocculi, but differing in the important fact that, whereas the calcium flocculi are bright, those of hydrogen are dark. This result was confirmed by subsequent photographs, and it was found that in general the hydrogen flocculi are dark, although in certain disturbed regions bright hydrogen flocculi appear. Some of these are eruptive in character, and correspond closely with the brilliant eruptive calcium flocculi. But in other cases, in regions where no violent eruptive disturbances seem to be present, the hydrogen flocculi frequently appear bright instead of dark. Such regions are usually in the immediate vicinity of active sun-spots, where it is probable that the temperature of the hydrogen is considerably higher than in the surrounding regions. Since a higher temperature would undoubtedly produce increased brightness, the spectroheliograph thus seems to afford a method of distinguishing between regions of higher and lower temperature—an additional property which should prove of great value in investigations on the vapors associated with sun-spots. It is possible, of course, that the increased brightness is due, not merely to an increase of temperature, but to other causes, perhaps of a chemical or electrical nature, which are not yet understood. But the assumption that increased temperature is the effective cause may be provisionally accepted as very probable.

The comparative darkness of the ordinary hydrogen flocculi evidently indicates that this gas in the flocculi for some reason radiates less light than the hydrogen gas which, probably after diffusing from the flocculi, has spread in a nearly uniform mass over the entire surface of the Sun. The simplest hypothesis is to assume that the diminished brightness of the flocculi is due to the reduced temperature in the upper chromosphere, where the absorption probably occurs. The results of work at Mount Wilson seem to render this view probable. It should be emphasized at this point, however, that the explanation of spectroheliograph results offered in this chapter is merely an hypothesis, which subsequent investigation may not prove to be correct. According to Julius, the flocculi are not luminous clouds, but the effects of anomalous dispersion of light passing out from the Sun's interior through vapors of unequal density.

The Rumford spectroheliograph was also used to secure photographs with some of the stronger dark lines of iron and other substances. But even with the grating the dispersion was insufficient to give thoroughly trustworthy results, except in a very few cases. It was evident that much greater dispersion must be employed in order to realize the full advantages of the method in future work.

Within a short time after the first work at the Kenwood Observatory the spectroheliograph came into general use. Evershed constructed and successfully used one of these instruments in England, and a year later Deslandres, whose admirable work on the spectra of the flocculi was contemporaneous with the investigations at the Kenwood Observatory, undertook systematic research with a spectroheliograph at the Paris Observatory. His contributions to the development of the instrument have been very valuable. Other spectroheliographs are now used daily in India, Sicily, Spain, Germany, England, and the United States.

24

Wave-Length Determinations and General Results obtained from a Detailed Examination of Spectra photographed at the Solar Eclipse of January 22, 1898†

J. EVERSHED

IN this paper the results are given of a detailed study and measurement of a series of spectra photographed at the eclipse of 1898, with a glass prismatic camera of $2\frac{1}{4}$ inches aperture. Ten exposures were made, all yielding good negatives, in which the great extension in the ultra-violet is a marked feature.

The first two photographs of the series were exposed at 20 seconds and 10 seconds before totality respectively, and are images of the cusp spectrum. They show the Fraunhofer lines with great distinctness, although the latter are much less dark than in the ordinary solar spectrum. The lines were measured and identified for the purpose of facilitating the reduction of the bright line spectra obtained during totality.

Spectrum No. 3 was exposed for 4 seconds, beginning 2 seconds before second contact. In this the flash spectrum is fully developed, and extends from λ 3340 to λ 6000. The majority of the bright arcs, including those due to the upper chromosphere, extend over 40° of the limb, implying a depth of $1''\cdot3$ for the gases composing this layer. The total depth of the chromosphere deduced from the hydrogen arcs is $8''\cdot2$, and from the calcium arcs $11''\cdot6$. There are 313 measurable lines in this negative.

† *Proc. Roy. Soc.* **68,** 6–9, 1901.

Spectrum No. 4, exposed for half a second shortly after second contact, gives the spectrum of the upper chromosphere and prominences. Seven of the latter are shown. The images are about equally dense in calcium radiations, although in hydrogen there is a marked variation of intensity between the different prominences.

A conspicuous feature in the spectrum of two of the prominences is a band of continuous spectrum, beginning at λ 3668 near the end of the hydrogen series, and extending indefinitely in the ultraviolet.

Good measures were obtained of the images of a small prominence at the centre of the plate.

Spectrum No. 5. This plate had a long exposure near mid-totality. The continuous spectrum of the corona is strongly marked, and the green corona line is well shown at position angles 60° to 78°, and 95° to 105°. A new corona line is faintly impressed at λ 3388 $\pm$, the maxima of intensity being at the same position angles as those of the green line.

Spectrum No. 7 shows the re-appearing arcs of the flash spectrum, the exposure ending about 4 seconds before third contact. The green corona line is shown on both east and west limbs, and there is a faint corona line near H.

Spectrum No. 8. This was exposed almost at the instant of third contact, the re-appearing photosphere showing as four narrow bands of continuous spectrum due to Baily's beads. The flash spectrum arcs extend between and across the bands, and can be traced over an arc of 55°, the depth of the layer, in this case exceeding 2″.

The focus in this negative is poor, and no measures were made; but as far as can be judged, comparing this plate and No. 3, the spectra of the east and west limbs of the sun are identical.

Spectra Nos. 9 *and* 10. These are cusp spectra, very similar to Nos. 1 and 2.

General Results and Conclusions

The flash spectrum. Comparing the wave-length values of the flash spectra with Rowland's wave-lengths of the solar lines, it is at once evident that practically all the strong dark solar lines are present in the flash as bright lines; and all the bright lines in the flash, excepting hydrogen and helium, coincide with dark lines having an intensity greater than three on Rowland's scale.

The relative intensities of the lines in the two spectra are, however, widely different, many conspicuous flash lines coinciding with weak solar lines, and some of the strong solar lines being represented by weak lines in the flash spectrum.

This, however, applies only to the spectrum taken as a whole. Selecting the lines of any one element, it is found that the relative intensities in the flash spectrum agree closely with those of the same element in the solar spectrum. This is particularly well shown in the case of the elements iron and titanium.

The want of agreement in the relative intensities of the lines of different elements in the bright line and dark line spectra is probably due to the unequal heights to which the various elements ascend in the chromosphere, a low-lying gas of great density giving strong absorption lines, but weak emission lines, on account of the excessively small angular width of the radiating area.

The more extensively diffused gases of small density, on the other hand, give strong emission lines in the flash spectrum, and weak absorption lines.

The spectrum arcs obtained with a prismatic camera are not true images of the strata producing them, but *diffraction* images more or less enlarged by photographic irradiation. Monochromatic radiations from a layer 2″ in depth will produce arcs or "lines" which are as narrow as can be defined by instruments of ordinary resolving power.

The intensities of these images do not represent the intrinsic intensities of the bright lines of the different elements; the apparent intensity of the radiation from an element depending on the extent of diffusion of that element above the photosphere.

But in the dark line spectrum the intensities depend on the total quantity of each absorbing gas above the photosphere irrespective of the state of diffusion of the different elements.

The flash spectrum as a whole appears from these results to represent the upper, more extensively diffused portion of a stratum of gas, which, by its absorption, gives the Fraunhofer spectrum.

Fifteen elements are recognised with certainty in the flash spectrum (No. 3), and five are doubtfully present. The atomic weights of these elements in no case exceed 91. All the known metals having atomic weights between 20 and 60 seem to be present in the lower chromosphere, but among these there does not seem to be any relation between the atomic weights and the elevations to which the gases ascend in the chromosphere.

The only non-metals found are H, He, C, and possibly Si.

Of the 225 lines measured in the ultra-violet region of the spectrum only 29 remain unidentified.

The hydrogen spectrum. Twenty-eight hydrogen lines are shown in spectrum No. 3. The wave-lengths obtained may be compared with the theoretical values derived from Balmer's formula. With the exception of Hδ, which seems to be unaccountably displaced towards the red, the wave-lengths of the ultraviolet lines are found to agree closely with the formula. A slight deviation occurs in the most refrangible lines, the positions of which seem to be distinctly more refrangible than those assigned by theory.

The continuous spectrum given by the prominences in the ultraviolet, beginning at the end of the hydrogen series, seems analogous to a feature noticed by Sir William Huggins in the absorption spectra of 1st type stars, and is possibly due to hydrogen.

Hydrogen and helium in the lower chromosphere. From the character of some of the helium lines it is inferred that this element is probably absent from the lowest strata, whilst parhelium appears to be separated from helium, and to exist at a lower level.

Unlike helium, hydrogen gives very intense lines in the flash layer. These lines are well defined and narrow, even in the very lowest strata.

Reasons can be given to show that the absence of hydrogen absorption in the ultraviolet, and of helium absorption in the visible spectrum, may be due to insufficient quantity of these elements above the photosphere, not to equality of temperature between the radiating gas and photospheric background.

The corona spectrum. The wave-length of the green line deduced from measures of No. 3 and No. 7 spectra confirms the value obtained by Sir Norman Lockyer at the same eclipse. The only other lines shown on these photographs are λ 3388 and near H.

25

On the Theoretical Temperature
of the Sun; under the Hypothesis
# of a Gaseous Mass maintaining its
Volume by its Internal Heat, and
depending on the Laws of Gases
as known to Terrestrial Experiment†

J. HOMER LANE

MANY years have passed since the suggestion was thrown out by Helmholtz, and afterwards by others, that the present volume of the sun is maintained by his internal heat, and may become less in time. Upon this hypothesis it was proposed to account for the renewal of the heat radiated from the sun, by means of the mechanical power of the sun's mass descending toward his center. Calculations made by Prof. Pierce, and I believe by others, have shown that this provides a supply of heat far greater than it is possible to attribute to the meteoric theory of Prof. Wm. Thomson, which, I understand, has been abandoned by Prof. Thomson himself as not reconcilable with astronomical facts. Some years ago the question occurred to me in connection with this theory of Helmholtz whether the entire mass of the sun might not be a mixture of transparent gases, and whether Herschel's clouds might not arise from the precipitation of some of these gases, say carbon, near the surface, with their revaporization when fallen or carried into the hotter subjacent layers of atmosphere beneath; the circulation necessary for the play of this Espian

† *Amer. Journ. Science and Arts* (2nd series), **50**, 57–74, 1870. Read before the National Academy of Sciences at the session of April 13–16, 1869.

theory being of course maintained by the constant disturbance of equilibrium due to the loss of heat by radiation from the precipitated clouds. Prof. Espy's theory of storms I first became acquainted with more than twenty years ago from lectures delivered by himself, and, original as I suppose it to be, and well supported as it is in the phenomena of terrestrial meteorology, I have long thought that Prof. Espy's labors deserve a more general recognition than they have received abroad. It is not surprising, therefore, in a time when the constitution of the sun was exciting so much discussion, that the above suggestions should have occurred to myself before I became aware of the very similar, and in the main identical, views of Prof. Faye, put forth in the *Comptes Rendus*. I sought to determine how far such a supposed constitution of the sun could be made to connect with the laws of the gases as known to us in terrestrial experiments at common temperatures. Some calculations based upon conjectures of the highest temperature and least density thought supposable at the sun's photosphere led me to the conclusion that it was extremely difficult, if not impossible, to make out the connection in a credible manner. Nevertheless, I mentioned my ideas to Prof. Henry, Secretary of the Smithsonian Institution, when he immediately referred me to a number of the *Comptes Rendus*, then recently received, containing Faye's exposition of his theory. Of course nothing is further from my purpose than to make any kind of claim to any thing in that publication. After becoming acquainted with his labors I still regarded the theory as seriously lacking, in its physical or mechanical aspect, the direct support of confirmatory observations, and even as being subject to grave difficulty in that direction. In this attitude I allowed the subject to rest until my friend Dr. Craig, in charge of the Chemical Laboratory of the Surgeon General's office, without any knowledge of Faye's memoir, or of my own suggestions previously made to Prof. Henry and another scientific friend, fell upon the same ideas of the sun's constitution, availing himself, precisely as I had done, of Espy's theory of storms. Dr. Craig's ideas were communicated to a company of scientific gentlemen early last spring, and soon after, Prof.

Newcomb, of the U.S. Naval Observatory, entered into a general survey of the nebular hypothesis. These communications of Dr. Craig and Prof. Newcomb led me to enter into a renewed examination of the mechanical embarrassment under which I had believed the theory to labor. Not any longer relying on my first rough estimate based on assumed high temperatures at the photosphere, the question was now inverted. Assuming the gaseous constitution, and assuming the laws expressed in Poisson's formulae, known to govern the constitution of gases at common temperatures and densities, what shall we find to be the temperatures and densities corresponding to the observed volume of the sun supposing it were composed of some known gas such as hydrogen, or supposing it to be composed of such a mixture of gases as would be represented by common air. Pure hydrogen will, of course, give us the lowest temperature of all known substances, under the general hypothesis.

The question was resolved, and the results were communicated in graphical and numerical form in May or June last to two or three scientific friends, but their publication has been delayed by an unavoidable absence of several months from home.

Premising that the unity of density shall correspond to a unit of mass in the cube of the unit of length, the unit of force to the force of terrestrial gravity in the unit of mass, and the unit of pressure or elasticity in the gas to the unit of force on a surface equal to the square of the unit of length:

Let $r=$ the distance of an element of the sun's mass from the sun's centre,

$t=$ the temperature of the element,

$\sigma t=$ its atmospheric subtangent, referred to the force of gravity at the earth's surface, or height of the column of homogeneous gas, whose terrestrial gravitating force would equal its elasticity,

$\rho=$ its density, or mass of its unit volume,

$=$ force of terrestrial gravity in its unit volume,

$\rho \sigma t=$ its elasticity, or elastic force per unit surface,

$R=$ the earth's radius,

M = the earth's mass,

m = the mass of the part of the sun's body contained in the concentric sphere whose radius is r,

$\dfrac{M\ r^2}{m\ R^2}\ \sigma\ t$ = the subtangent of the gas under its actual gravitating force in the sun.

The condition of equilibrium between the gravitating force of a thin horizontal layer of gas whose thickness is dr, and the difference of elastic force between its lower and upper surfaces, is expressed by the equation,

$$d\,\rho\,\sigma\,t = -\frac{m\ R^2}{M\ r^2}\,\rho\,d\,r.$$

Under the hypothesis that the law of Mariotte and the law of Poisson prevail throughout the whole mass, and that this mass is in convective equilibrium, we have

$$\sigma = \text{a constant,} \qquad (1)$$
$$t = t_1\,\rho^{k-1},\dagger$$

t_1 representing the value of t in the part of the mass where the density is a unit.

The theoretical difficulties which, if the supply of solar heat is to be kept up by the potential due to the mutual approach of the parts of the sun's mass consequent on the loss of heat by radiation, come in when we suppose a material departure from these laws of Mariotte and of Poisson at the extreme temperatures and pressures in the sun's body, or how far such difficulties intervene, will be considered further on.

By means of the constant value of σ, and the value of t given in (1), the above differential equation is transformed into

$$k\,\sigma\,t_1\,\rho^{k-2}d\rho = -\frac{m\ R^2}{M\ r^2}\,dr,$$

the integral of which gives

$$1 - \left(\frac{\rho}{\rho_0}\right)^{k-1} = \frac{k-1}{k}\,\frac{R^2}{\sigma\ Mt_1\,\rho_0{}^{k-1}}\int_0^r \frac{mdr}{r^2}, \qquad (2)$$

† k represents the ratio of the specific heat of a gas under constant pressure to its specific heat under constant volume.

in which ρ_0 is the value of ρ at the sun's centre.

We have also

$$m = 4\pi \int_0^r \rho r^2 dr = 4\pi \rho_0 \int_0^r \frac{\rho}{\rho_0} r^2 dr. \tag{3}$$

If now we put

$$r = \left(\frac{k\sigma \, Mt_1}{4(k-1)R^2 \pi \rho_0^{2-k}} \right)^{\frac{1}{2}} x, \tag{4}$$

we shall have

$$m = 4\pi \rho_0 \left(\frac{k\sigma \, Mt_1}{4(k-1)R^2 \pi \rho_0^{2-k}} \right)^{\frac{3}{2}} \mu, \tag{5}$$

in which

$$\mu = \int_0^x \frac{\rho}{\rho_0} x^2 dx, \tag{6}$$

and equation (2) becomes

$$1 - \left(\frac{\rho}{\rho_0} \right)^{k-1} = \int \frac{\mu dx}{x^2}. \tag{7}$$

In equations (6) and (7) it is plain that upon the value of k alone depends: first the form of the curve that expresses the value of ρ/ρ_0 for each value of x; secondly, the value of the upper limit of x corresponding to $\rho/\rho_0 = 0$; and thirdly, the corresponding value of μ. These limiting, or terminal, values of x and μ, cannot be found except by calculating the curve, for equations (6) and (7) seem incapable of being reduced to a complete general integral. But when these values have been found for any proposed value of k, they may be introduced once for all into equations (4) and (5), from which the values of ρ_0, and of σt_1, are at once deduced.

I have made these calculations for two different assumed values of k, viz., $k = 1\cdot4$, which is near the experimental value it has in common air, and $k = 1\frac{2}{3}$, which is the maximum possible value it can have in the light of Clausius' theory of the constitution of the gases. The calculation of the curve of ρ/ρ_0, or of $(\rho/\rho_0)^{k-1}$, begins at the sun's centre where $x = 0$. For the small values of x, integration by series enables us readily to deduce from equations (6) and (7) the following approximate numerical equations:

$$\text{For } k = 1\cdot4,$$

$$\mu = \tfrac{1}{8}x^3 - \tfrac{1}{12}x^5 + \tfrac{5}{336}x^7 - \tfrac{125}{54432}x^9 + \&\text{c.} \qquad (8)$$

$$1 - \left(\frac{1}{\rho_0}\right)^{0\cdot4} = \tfrac{1}{6}x^2 - \tfrac{1}{48}x^4 + \tfrac{5}{2016}x^6 - \tfrac{125}{435456}x^8 + \&\text{c.} \qquad (9)$$

$$\text{For } k = 1\tfrac{2}{3},$$

$$\mu = \tfrac{1}{3}x^3 - \tfrac{1}{20}x^5 + \tfrac{1}{240}x^7 - \tfrac{1}{3888}x^9 + \&\text{c.} \qquad (10)$$

$$1 - \left(\frac{\rho}{\rho_0}\right)^{\tfrac{2}{3}} = \tfrac{1}{6}x^2 - \tfrac{1}{80}x^4 + \tfrac{1}{1440}x^6 - \tfrac{1}{31104}x^8 + \&\text{c.} \qquad (11)$$

For larger values of x, until $(\rho/\rho_0)^{k-1}$ becomes sufficiently small as there is no need of great precision in these calculations, I have merely developed the values of μ and $(\rho/\rho_0)^{k-1}$ corresponding to $x=1\cdot1$, $x=1\cdot2$, $x=1\cdot3$, etc., by means of differences taken from the differential coefficients at the middle of each increment of x, and for the same reason have thought it sufficient to begin with $x=1$, in equations (8) and (9) or (10) and (11). After arriving at a sufficiently small value of $(\rho/\rho_0)^{k-1}$ the calculation is finished by aid of the following approximate equations also derived by integration from (6) and (7).

$$\mu' - \mu = \frac{k-1}{k}\mu'^{\frac{1}{k-1}}x'^{2-\frac{2}{k-1}}(x'-x)^{1+\frac{1}{k-1}}(1+X) \qquad (12)$$

$$\left(\frac{\rho}{\rho_0}\right)^{k-1} = \frac{\mu'(x'-x)}{x'x} - \frac{(k-1)^2}{k(2k-1)}\mu'^{\frac{1}{k-1}}x^{-\frac{2}{k-1}}(x'-x)^{2+\frac{1}{k-1}}$$

$$-\frac{(k-1)(2k^2-3k+2)}{k(2k-1)(3k-2)}\mu'^{\frac{1}{k-1}}x'^{-1-\frac{2}{k-1}}(x'-x)^{3+\frac{1}{k-1}} \qquad (13)$$

In these equations x' and μ' are the values of x and μ corresponding to $\rho/\rho_0 = 0$, or the upper limit of the supposed solar atmosphere, and

$$X = -\frac{k(2k-3)}{(k-1)(2k-1)}\frac{x'-x}{x'} + \frac{k(k-2)(2k-3)}{2(k-1)^2(3k-2)}\frac{(x'-x)^2}{x'^2} + \text{etc.}$$

$$-\frac{k-1}{2k(2k-1)}\mu'^{-1+\frac{1}{k-1}}x'^{2-\frac{2}{k-1}}(x'-x)^{1+\frac{1}{k-1}} + \text{etc.}$$

With the values of x' and μ' determined, using r' and m' to express in like manner the corresponding values of r and m at the upper limit of the theoretical atmosphere, we find from equations (4) and (5)

$$\rho_0 = \frac{m'x'^3}{4\pi\mu'r'^3}, \tag{14}$$

$$\sigma t_1 = \frac{4\pi(k-1)R^2r'^2\rho_0^{2-k}}{kMx'^2},$$

and by equation (1), $\sigma t = \dfrac{4\pi(k-1)R^2r'^2\rho_0}{kMx'^2}\left(\dfrac{\rho}{\rho_0}\right)^{k-1}$, \hfill (15)

$$= \frac{k-1}{k}\frac{m'R^2x'}{\mu'Mr'}\left(\frac{\rho}{\rho_0}\right)^{k-1} \tag{16}$$

A glance at equation (7) will show that $\mu'(x'-x)/x'x$, equation (13), or $(\mu'/x')\,(r'-r/r)$ may be taken equal to $(\rho/\rho_0)^{k-1}$ throughout the considerable upper part of the volume of the hypothetic gaseous body in which $1-(\mu/\mu')$, or $(1-m/m')$, is sufficiently small to be neglected. This substitution in the last equation gives

$$\sigma t = \frac{k-1}{k}\frac{m'R^2}{Mrr'}(r'-r), \text{ nearly,} \tag{17}$$

and also $\quad \rho = \left(\dfrac{\mu'}{x'}\right)^{\frac{1}{k-1}}\rho_0\left(\dfrac{r'-r}{r}\right)^{\frac{4}{k-1}}$ nearly,

$$= \frac{1}{4\pi}\mu'^{-1+\frac{1}{k-1}}x'^{3-\frac{1}{k-1}}\frac{m'}{r'^3}\left(\frac{r'-r}{r}\right)^{\frac{1}{k-1}} \tag{18}$$

Now the mechanical equivalent of the heat in the mass ρ of a cubic unit in volume of any perfect gas whose atmospheric subtangent is σt, is $[1/(k-1)]\,\rho\cdot\sigma t$, and the mechanical equivalent of the heat that it would give out, in being cooled down under constant pressure to absolute zero, is $[k/(k-1)]\,\rho\cdot\sigma t$. If the density ρ is taken in units of the density of water, and the unit of length be the foot, this expression is multiplied by $62\frac{1}{2}$ to give for the mechanical equivalent in foot pounds

$$62\frac{1}{2}\frac{k}{k-1}\rho\cdot\sigma t = \frac{62\frac{1}{2}}{4\pi}\mu'^{-1+\frac{1}{k-1}}x'^{3-\frac{1}{k-1}}\frac{m'^2R^2}{Mr'^4}\left(\frac{r'-r}{r}\right)^{1+\frac{1}{k-1}} \quad (19)$$

The mechanical equivalent $[1/(k-1)]\,\rho\cdot\sigma t$, of the heat in the mass ρ, viewed in the light of Clausius' mechanical theory of the gases, includes the motions of the separate atoms of each supposed compound molecule relatively to each other, as well as the motion of translation which each compound molecule makes in a straight path through free space till it impinges upon another compound molecule. If we wish to find the mechanical equivalent which would be due to this motion of translation alone, we must put $k=1\frac{2}{3}$ in the factor $1/(k-1)$ by which $\rho\cdot\sigma t$ is multiplied, and this gives $\frac{2}{3}\rho\cdot\sigma t$. To find from this the mean of the squares of the velocities of translation of the compound molecules, we divide by the mass ρ, and, if the foot be the unit of length, multiply by $64\cdot3$, whence we have for the velocity found by taking the square root of this mean of the squares

$$8\cdot02\sqrt{\tfrac{3}{2}\sigma t} = 8\cdot02\left(\frac{3}{2}\frac{k-1}{k}\frac{m'R^2x'}{\mu'Mr'}\right)^{\frac{1}{2}}\left(\frac{\rho}{\rho_0}\right)^{\frac{k-1}{2}} \quad (20)$$

Determination of the curve of density for $k=1\cdot4$. Beginning with $x=1$, in equations (8) and (9), we find $\mu=0.2626$ and $(\rho/\rho_0)^{\frac{1}{16}}=0\cdot8520$. Developing the values of μ and $(\rho/\rho_0)^{\frac{1}{16}}$ for $x=1\cdot1$, $x=1\cdot2$, etc., by means of differences we arrive at the values $\mu=2.145$ and $(\rho/\rho_0)^{\frac{1}{16}}=0\cdot1378$ when $x=4\cdot0$. Putting these values into equations (12) and (13) we find

$$x'=5.355, \ \mu'=2\cdot188$$

If we now allow $\frac{1}{22}$d of the radius of the photosphere, or about 20,000 miles, for the height of the theoretic upper limit of the solar atmosphere above the photosphere, and if we take the mean specific gravity of the earth's mass at $5\frac{1}{2}$, and the mean specific gravity of the sun within the photosphere at $\frac{1}{4}$ that of the earth, as it is known to be, these values of x' and μ' give us in equation (14)

$$\rho_0 = 28\cdot16,$$

so that the density of the sun's mass at the center would be nearly one-third greater than that of the metal platinum.

Curve of density for $k = 1\frac{2}{3}$. For this value of k the numerical coefficients in equations (8) and (9) are replaced by those in (10) and (11). Otherwise, the same process employed with the value $k = 1\cdot4$, gives, starting with $x = 1$, $\mu = 0\cdot2875$ and $(\rho/\rho_0)^{\frac{3}{2}} = 0\cdot8452$, and developing for $x = 1\cdot1$, $x = 1\cdot2$, etc., brings us to $\mu = 2\cdot557$ and $(\rho/\rho_0)^{\frac{3}{2}} = 0\cdot1591$, for $x = 3\cdot0$, and finally gives us

$$x' = 3\cdot656, \ \mu' = 2\cdot741,$$

and if we now assume the same height as before for the theoretic upper limit of the sun's atmosphere, instead of $\rho_0 = 28\cdot16$, we find

$$\rho_0 = 7\cdot11.$$

The new curve of density is found in the same way as the first, and is presented to the eye in the diagram in comparison with it. In the upper part of both curves the scale of density is increased ten-fold, and it is, in part only, evident to the eye how immensely different, for the two values of k, becomes the density in the upper parts of the sun's mass. It appears to the eye only in part because the ratio of the two densities multiplies itself rapidly in approaching the upper limit of the atmosphere.

The above was communicated in writing as here given, to the Academy at its late session.† The draft of the following, and a part of the details of its substance, have been prepared since.

† I desire here to state that the formulae which show the relation between the temperature, the pressure, the density, and the depth below the upper limit of the atmosphere, so far as they apply to the upper part of the sun's body, were independently pointed out by Prof. Peirce, in a very interesting paper which that distinguished physicist read before the Academy at the same session, and prior to the presentation of this paper. Also to recall a fact

Equation (20) gives in feet the square root of the mean square of velocity of translation of molecules $(8 \cdot 02\sqrt{\tfrac{2}{3}}\sigma t)$. At the sun's center we find this would be 331 miles per second for the curve of density corresponding to $k = 1\tfrac{2}{3}$, and 380 miles per second for the curve of density corresponding to $k = 1 \cdot 4$.

In 1838 Pouillet, following the law of heat radiation given by Dulong and Petit, estimated the temperature of the radiating surface of the sun, from observations by himself of the quantity of heat it emits, at from 1461°C. to 1761°C. Herschel, from Pouillet's observations, and his own made at the Cape of Good Hope about the same time, adopts, after allowing one-third for the absorption of our atmosphere, forty feet as the thickness of ice that would be melted per minute at the sun's surface. The temperature of the radiating surface calculated from this datum by the formula of Dulong and Petit, and with the co-efficient of radiation found by Prof. W. Hopkins for sandstone, the smallest co-efficient he found, is 1550° C. or 2820° F. But then the solar radiation is many thousands of times greater than the greatest in Dulong and Petit's experiments, so that these calculations of the temperature of the sun's photosphere have little weight notwithstanding the simplicity and accuracy with which the formula represents the experiments from which it was derived. Nothing authorizes us to accept the formula as more than an empirical one. It seems desirable that experiments similar to those of Dulong and Petit should be made on the rate of cooling of intensely heated bodies, such as balls of platinum not too large. By placing the heated ball in the centre of a hollow spherical jacket of water, either flowing or in an unchanged mass, the quantities of heat radiated in successive equal spaces of time will be determined, and the corresponding differences of temperature in the heated ball can at least be estimated with whatever probability we may

---

which I first learned from Prof. Peirce's mention of it to the Academy, viz. that Prof. Henry long ago threw out the idea of the atmospheric condition to which Prof. Thomson has more recently given the term convective equilibrium, viz. such that any portion of the air, on being conveyed into any new layer above or below, would find itself reduced, by its expansion or compression, to the temperature of the new layer.

rely on our knowledge of the specific heat of its material. At present the best means we have of forming any judgement of the probable temperature of the source of the sun's radiation, is perhaps to be found in a comparison between the effects of the hydro-oxygen blowpipe, and the recorded effects of Parker's great burning lens. I am not aware that this method has before been resorted to.

If the angle of aperture at the focus of a burning lens, or combination of lenses, be called $2a$, the radiation received by a small flat surface at the focus will be $\sin^2 a$, if a unit be taken to represent the radiation the same small flat surface would receive just at the sun's surface. Parker's lens, with the small lens added, had, at the focus so formed, an angle of aperture of about 47°. A small flat surface at its focus would therefore receive about one-sixth the radiation that it would just at the sun, making no allowance for absorption by the atmospheres of the earth and sun and rays lost in transmission through the lenses. Pouillet, from the experiments already alluded to made by himself, found his atmosphere in fine weather transmitted, of the sun's heat rays, about the fraction $\frac{4}{5}$ raised to a power whose exponent is the secant of the sun's zenith distance. This, of course, leaves out of view the heat rays of low intensity which are totally absorbed by the atmosphere. He also concluded from comparison with other experiments of his own with a moderately large burning glass, that that glass transmitted $\frac{7}{8}$ of the heat rays incident on it. If we assume the same fraction for each of the two lenses of Parker's combination, and assume further that the sun's zenith distance did not exceed 48° in the experiments made with it, we find for the fractional multiplier expressing the part of the sun's heat radiation which arrived at the focus unintercepted, $(\frac{4}{5})^{1 \cdot 2}(\frac{7}{8})^2 = \cdot 55$. Hence the radiation actually received by a small flat surface at the focus was 0·09, or about one-eleventh, of what it would receive just at the sun. The heat so received by any body so placed in the focus, must, after the body has acquired its highest temperature, be emitted from it at the same rate. The heat so emitted will consist: first, of heat radiated into that part of space toward which the radiating surface of the body looks; secondly, of heat carried off by convection of the air; thirdly, of heat conducted away by the body supporting the body subjected to experiment; fourthly, of heat rays, if any, reflected, and not absorbed, by the body subjected to experiment. Assuming it as a reasonable conjecture that full half of all this† consists of heat *radiated* into the single *hemisphere* looking upon a flat surface, we may conclude that the body, at its highest acquired temperature, radiated not less than $\frac{1}{20}$th as much heat as is radiated by an equal extent of surface of the sun's photosphere, over and above such part of that radiation as may be intercepted by the sun's atmosphere, and such rays of low intensity as are *totally* absorbed by our own atmosphere, the whole of which apparently cannot be great. No allowance seems necessary for the chromatic and spherical dispersion of the lenses, since the diameter of the focus is stated at half an inch, while the true diameter of the sun's image would be not less than one-third of an inch.

Now we are not without the means of forming a probable approximate estimate of this temperature at which the radiation becomes $\frac{1}{20}$th, more or less, of that of the sun's photosphere. We are told that in the focus of Parker's compound lens 10 grains of very pure lime ("white rhomboidal spar") were melted in 60 seconds. We may presume that in that length of time the temperature of the lime, after parting with its carbonic acid, made a near approximation to the maximum at which it would be stationary, a presumption confirmed by the period of 75 seconds said to have been occupied in the fusion of 10 grains of carnelian, and by the considerable period of 45 seconds for the fusion of a topaz of only 3 grains, and 25 seconds for an oriental emerald of but 2 grains, and in fact sufficiently proved, it would seem, by

† As to the heat carried off by convection of the air, if its quantity be calculated by the formula given by Dulong and Petit for that purpose, it comes out utterly insignificant in comparison with the heat received from the burning glass. The conjectural allowance of $\frac{4}{9}$ths in all, of this, is likely, therefore, to be much too large. Not much reliance, indeed, can be placed upon the formula here mentioned, at such a temperature as 4000° F., yet, as by it the convection is taken proportional to the 1·233 power of the difference of temperature, it seems unlikely that it gives a quantity very many fold less than the truth.

observing that the heat we have estimated to fall at the focus, upon a flat surface, would suffice, if retained, to raise the temperature of a quarter of an inch thick of lime 4000° F. in 5 seconds. If, then, we may take the temperature *maintained* at the focus of Parker's lens to have been at the melting point of lime, we may conclude that it is also not far from the temperature given by the hydro-oxygen blowpipe. Dr. Hare, who was the first inventor of this instrument, and the discoverer of its great power, melted down, by its means, in partial fusion, a very small stick of lime cut on a lump of that material, which we understand to have been a very pure specimen. Burning glass and blowpipe seem each to have been near the limit of its power in this apparently common effect. But Deville found the temperature produced by the combination of hydrogen and oxygen under the atmospheric pressure to be 2500° C. As the lime in the heated blast would radiate rapidly, its temperature must have been lower than that of combined hydrogen and oxygen, and I have called it 2220° C. or 4000° F.

The formula of Dulong and Petit, with the coefficient found by Hopkins, as already mentioned, gives for the quantity of heat radiated in one minute by a square foot of surface of a body whose temperature is $\theta + t$ centigrade, into a chamber whose temperature is θ centigrade, when expressed with the unit employed by Hopkins,

$$8 \cdot 377 \ (1 \cdot 0077)^{\theta} \ [(1 \cdot 0077)^{t} - 1].$$

It will be convenient, and, in the discussion of the high temperatures with which we are concerned, will involve no sensible error, to use the hypothesis that the space around the radiating body is at the temperature of 0° C. and the formula for the radiation then becomes,

$$8 \cdot 377 \ [(1 \cdot 0077)^{t} - 1]. \tag{21}$$

The unit used by Hopkins, in the formula here given, is the quantity of heat that will raise the temperature of 1000 grains of water 1° C. Expressed by the same unit, the quantity adopted by

Sir J. Herschel as the amount of the sun's radiation, viz. that which would melt 40 feet thick of ice in a minute (at the sun's surface), is 1,280,000. The $\frac{1}{20}$th of this, or 64,000, expresses, therefore, the quantity which we have estimated the lime under Parker's lens to have radiated, per square foot of its surface, at its estimated temperature of 4000° F. If now we calculate its temperature by the above formula, from the estimated radiation, the result is 1166° C. or 2130° F. This is manifestly much below the real temperature, and so far below that there can be no doubt the formula of Dulong and Petit has failed at the melting point of lime. If instead of the coefficient 8·377 we had used the larger coefficient 12·808 which Hopkins gives for unpolished limestone, the formula would have been reduced only 53° C. It best suits the direction of our inquiry to use the smallest coefficient which Hopkins' experiments gave, since we are seeking the *highest* temperature which can be plausibly deduced from the sun's radiation. For ease of expression, the curve which we will imagine for representing the actual relation of radiation to temperature, the horizontal ordinate standing for the temperature and the vertical ordinate for the radiation corresponding thereto, may be called the curve of radiation. The course of this curve from the freezing point of water to a point somewhat below the boiling point of mercury is correctly marked out to us by the formula. Beyond that we have but the rough approximation which we can get by means of the above comparison, to the single point of the curve where the radiation is $\frac{1}{20}$th that of the sun's photosphere. The attempt, from these data, to extend the curve till it reaches the full radiation of that photosphere, must be mainly conjectural. As a basis for the most plausible conjecture I am able to make let us assume: first, that the upward concavity of the curve of radiation, which increases very rapidly with the temperature as far as the curve follows the formula of Dulong and Petit, is at no temperature greater than that formula would give it at the same temperature; secondly, that the curve of radiation is nowhere convex upward. If, then, we set out from these two conjectural assumptions—of the degree of probability of which each one must form his own impression—the greatest temperature the sun's photosphere could have consistently with the radiation of 64,000 at the temperature of 4000° F., is found by drawing through the point representing that radiation and that temperature a straight line tangent to the curve of the formula. The line so drawn would cross the real curve of radiation in a greater or less angle at the radiation of 64,000 and temperature of 4000° F., and at higher temperatures would fall more or less *below* that curve, and its intersection with the sun's radiation of 1,280,000 would be at a temperature greater than that of the curve, that is to say, greater than the temperature of the sun's photosphere. This greater temperature is 55,450° F.

A different train of conjecture led me at first to assume a temperature of 54,000° F., and this last number I will here retain since it has been already used as the basis of some of the calculations we now proceed to give. It must be here recollected that we are discussing the question of *clouds* of *solid* or at least *fluid* particles floating in non-radiant gas, and constituting the sun's photosphere. If the amount of *radiation* would lead us to limit the temperature of such clouds of solids or fluids, so also it seems difficult to credit the *existence* in the solid or fluid form, at a higher temperature than 54,000° F. of any substance that we know of.

If then we suppose a temperature of 54,000° F., what would be the density of that layer of the hypothetic gaseous body which has that temperature, and what length of time would be required, at the observed rate of solar radiation, for the emission of all the heat that a foot thick of that layer would give out in cooling down under pressure to absolute zero? The latter question depends on the mechanical equivalent of this heat for a cubic foot of the layer of gas, and the two questions, together with that of the depth at which the layer would be situated below the theoretic upper limit of the atmosphere, are answered by equations (17), (18), and (19), provided we knew the value of k and the value of σ in the body of gas. The less the atomic weight of the gas the greater the value of σ, and the greater the density of the layer of 54,000° F. and the greater the quantity of heat which a cubic foot of it would give out in cooling down. I therefore base the first calculation on hydrogen as it is known to us. The value of σ is in that case about 800 feet, and the value of k about 1·4, nearly the same as in common air. These values would give for the layer of 54,000° F. a specific gravity about 0·00000095 that of water, or about $\frac{1}{90}$th that of hydrogen gas at common temperature and pressure, and the mechanical equivalent of the heat that a cubic foot of the layer would give out in cooling down under pressure to absolute zero would be only about 9000 foot-pounds, whereas the mechanical equivalent of the heat radiated by one square foot of the sun's surface in one minute is about 254,000,000 foot-pounds. The heat emitted each minute would, therefore, be fully half of all that a layer ten miles thick would give out in cooling down to zero, and a circulation that would dispose of volumes of cooled atmosphere at such a rate seems inconceivable.

It may possibly appear to some minds that the difficulty presented by this aspect of the case will vanish if we suppose the photosphere, or its cloudy particles, to be maintained by radiation at a temperature to almost any extent lower than that of convective equilibrium. This would enable us to place the theater of operations in a lower and denser layer of atmosphere, but the supposition seems to me difficult to realize unless, as the hot gases rise from beneath, precipitation could commence at a temperature many times higher than the 54,000° F. which we have estimated for the upper visible surface of the clouds, and this, as before intimated, seems to me itself extremely improbable.

I may mention here that my friend Dr. Craig, in an unpublished paper, following the hint thrown out by Frankland, is disposed to favor the idea that the sun's radiation may be the radiation of hot gases instead of clouds. At present I shall offer no opinion on that point one way or the other, but will only state it as my impression that if the theory of precipitated clouds, as above presented, is the true one, something quite unlike our present experimental knowledge, or at least much beyond it, is needed to make it intelligible.

The first hypothesis which offers itself in an attempt to make the theory rational is suggested by one point in Clausius' theory of the constitution of the gases, already alluded to. In forming his theory Clausius found that the known specific heats of the gases are all much too great for free simple atoms impinging on one another, and he therefore introduced the hypothesis of compound molecules, each compound molecule being a system of atoms oscillating among each other under forces of mutual attraction. Now if this were accepted as the actual constitution of the gases it is of course easy enough to conceive that in the fierce collisions of these compound molecules with each other at the temperatures supposed to exist in the sun's body, their component atoms might be torn asunder, and might thenceforth move as free simple molecules. In this case, still retaining the hypothesis of Clausius' theory, that the average length of the path described by each between collisions is large compared with the diameter of the sphere of effective attraction or repulsion of atom for atom, the value of k would reach its maximum of $1\frac{2}{3}$. Experiment has not shown us any gas in this condition, and for the present it is hypothetical. Even in hydrogen the value of k does not materially, if any, exceed the value of $1\cdot4$ which it has in air. But if it were found that the hydrogen molecule is compound, and that in the body of the sun the heat splits this molecule into two equal simple atoms, and in fact that all the matter in the sun's body is split into simple free atoms equally as small, then, while the value of k would be $1\frac{2}{3}$, the value of σ would be about 1600 feet. If with these values we repeat the calculation of the density of the layer of 54,000° F. we find its specific gravity to be 0·000363 of that of water, or 4·35 times that of hydrogen gas at common temperature and pressure and in its known condition, or 8·7 times that which the hydrogen in the hypothetic condition would have if it retained that condition at common temperature and pressure. We find also that the mechanical equivalent of all the heat that a cubic foot of the layer would give out in cooling down, under pressure, to zero, would be no less than 13,500,000 foot-pounds. Instead, therefore, of a layer ten miles thick, it would now require only a thickness of 38 feet to give out, in cooling down to zero, twice the heat emitted by the sun in one minute. It will be seen (equations (17) and (19)) that this thickness, retaining the constant value $k = 1\frac{2}{3}$, would diminish with the $2\frac{1}{2}$ power of the masses of the atoms into which the sun's body is hypothetically resolved (the reciprocal of the value of σ), and I leave each to form his own impression how far this view leads towards verisimilitude.

It is important to add that the depth of the layer of 54,000° F. below the theoretic upper limit of atmosphere, when calculated with value $k = 1\cdot4$, $\sigma = 800$ feet, comes out only 1107 miles, and with the values $k = 1\frac{2}{3}$ and $\sigma = 1600$ feet only 1581 miles. This calculation of the depth, unlike the other results above, may be said to be independent of the question of the constitution of the sun's interior mass. It is alike difficult, on any plausible hypothesis, to reconcile a temperature no higher than 54,000° F. with any perceptible atmosphere extending many thousand miles above, and yet no less an authority than Prof. Peirce has assigned a hundred thousand miles as the height of the solar atmosphere above the photosphere, at the same time, however, pointing out the enormous temperature which, under convective equilibrium, this would imply at the level of the photosphere. But all are not yet agreed that the appearances seen at such distances from the sun are proof of the existence of a true atmosphere there. It will be seen that the numbers I give above were obtained from a first hypothesis of an atmospheric limit 20,000 miles above the photosphere, but for the purpose of this paper it is of no consequence to repeat the calculation from a different limit.

It is, I believe, recognized on theoretical grounds that in an atmosphere containing a mixture of gases of unequal density the lighter gases might be expected to diffuse in greater proportion into the higher parts of the atmosphere and the heavier gases into the lower parts. But perhaps the supposed circulation which the emission of heat maintains within the photosphere must renew mixture at a rate sufficient to mask the rate which theory would assign for diffusion. I have not attempted a theoretic comparison between these two tendencies. It will suffice here to repeat that the above numerical results, so far as they may be thought to give countenance to the theory in its mechanical aspect, require that the entire inner mass of the sun shall have, at a mean (in the supposed state of dissociation), the very small atomic weight specified. We may notice in this connection the uniform proportion of oxygen and nitrogen gases in our atmosphere at the height of four miles or more at which the analysis has been made. Without having gone into a critical examination of the question, I suppose that at that height the proportion of oxygen which the theory of diffusive equilibrium would assign is notably diminished, and that it would be found that the circulation of the air is sufficiently active to mask the theoretic rate of diffusion.

The second hypothesis which might offer itself in an attempt to make the theory rational, but which a very little reflection is, I think, sufficient to set aside, is that which would modify Clausius' theory of the gases by assuming that in the sun's body the average length of the excursion made by each molecule between two consecutive collisions, becomes very short compared with the radius of the sphere of repulsion of molecule for molecule, and with the average distance of their centers at nearest approach. This way of harmonizing the actual volume of the sun with such a temperature as 54,000° F. in the photosphere, and with the smallest density which we can credit in the photosphere, would involve the consequence that the existing density of almost the entire mass of the sun is very nearly uniform and at its maximum possible, or at all events that any further sensible amount of collapse must be productive of but a very small amount, comparatively, of renewed supplies of heat, for the obvious reason that this hypothesis carries with it almost the entire neutralization of the force of gravity by the forces of molecular repulsion. In like manner it involves the consequence that in any such small contraction of the photosphere as can have taken place within the history of total eclipses, it is but a very small fraction of the sun's mass, near its surface, that can have taken part in the collapse to any thing like a proportionate extent. Hence it also extremely restricts the period during which we could suppose the sun to have existed under anything like its present visible magnitude in the past, consistently with the production in the way supposed of the supplies of heat it has been sending out. Another thing involved in this second hypothesis is the fact which Prof. Peirce has pointed out to the Academy, viz. that the existing molecular repulsion in the sun's body would immensely exceed such as would be indicated by the modulus of elasticity of any form of matter known to us.

In conclusion, I do not mean to say that there is any *invincible logical exclusion* of any law of the action of gases different from what is specified or alluded to above. I only mean that, so far as I can see, any theory of heat which is based *simply* and *solely* upon *molecular attraction* and *repulsion* dependent on molecular distance alone, cannot in its application to the sun, escape from the conditions indicated in this paper. It is certainly not absurd to imagine heat to be an agent of some kind so constituted that it cannot be thus represented by the sole conditions of motion and of molecular attraction and repulsion, but yet so constituted that in its effects upon matter it follows the conditions of mechanical equivalency as defined by Joule. In fact, such exceptional cases as the expansion of water in freezing seem to favor such a view, though the range of that phenomenon is very limited. One way of forming a mechanical representation of such a constitution would be by associating molecular motion with the mechanical powers, either with or without molecular attraction or repulsion; the manner in which the imagined mechanical power (or link) attaches itself to the molecules which it connects—so as to make their motion determine their mutual approach or recession or change of relative direction—being dependent on the existing motions and other conditions in such a way as to produce the observed phenomena. The possibility of such a mechanical *representation* is sufficient to show that such a supposed *constitution* is not *logically excluded*, but to accept such a mechanical representation as a physical fact is quite another matter, and, as it seems to me, a very difficult one. Of course this difficulty does not present itself when we suppose that heat is not motion.

26

Concerning the Equilibrum of the Solar Atmosphere†

K. Schwarzschild

Presented at the Session of 13 *January* 1906

1. Summary

THE solar surface shows us varying states and violent changes in its granulation, sunspots and prominences. In order to grasp the physical situation in which these phenomena occur, the spatial and temporal variations are usually as a first approximation replaced by an average stationary state, a mechanical equilibrium of the solar surface. Generally the stress has hitherto been laid on the so-called *adiabatic* equilibrium as prevailing in the atmosphere of the Earth when it is thoroughly mixed by ascending and descending currents. Here I would like to draw attention to equilibrium of another kind which one can call "radiative equilibrium". Radiative equilibrium will arise in a strongly radiating and absorbing atmosphere in which the mixing effect of the ascending and descending currents is unimportant in comparison with the heat exchange through radiation. Whether adiabatic or radiative equilibrium is more applicable to the sun, would be difficult to decide on general grounds. However, there are observational data which may lead to a possible decision. The solar disc is not uniformly bright, but gets darker from the centre to the limb. We can from this brightness distribution on the surface, by making some plausible assumptions, deduce the temperature distribution in the interior. The result is that the equilibrium of the solar atmosphere corresponds by and large to radiative equilibrium.

† Nachrichten von der Königlichen Gesellschaft der Wissenschaften zu Göttingen, 1906, p. 41.

The considerations which lead to this result assume that Kirchhoff's law applies—in other words—that the radiation of the solar atmosphere is pure thermal radiation. They further assume that upon penetrating into the solar interior one meets with a continuous change of state and does not change discontinuously from a fairly transparent chromosphere into a non-transparent photosphere formed from luminous clouds. One neglects the effect of the scattering of the light through diffraction at particles in the atmosphere, to the significance of which A. Schuster[†] has drawn attention, and also one neglects the refraction which von Seeliger[‡] invokes in order to explain the observed brightness distribution. One also neglects the different absorption of different wavelengths, the decrease of gravity with altitude, and the spherical propagation of the radiation. *The whole discussion can in no way therefore be looked upon as final or conclusive, yet it may provide a basis for further speculations in that a simple idea in the first instance is developed in its simplest form.*

2. Different Kinds of Equilibrium

One denotes the pressure by p, the absolute temperature (in degrees centigrade) by t, the density by ρ, molecular weight (relative to the hydrogen atom) by M, gravity by g, and the depth in the atmosphere (as reckoned inwardly from an arbitrary starting point) by h. The units are taken from the conditions at the surface of Earth, so that the unit of p will be an atmosphere (atm), the unit of ρ the air density at $273°$ absolute temperature under a pressure of 1 atm, the unit of g is the gravity on the surface of Earth, and the unit of h is the altitude of the so-called "homogeneous atmosphere" which is 8 km.

The following relation then holds good for an ideal gas:

$$\rho t = \frac{p.M}{R} \quad (R = 0{\cdot}106) \tag{1}$$

[†] *Astrophysical Journal*, 1905, **21**, p. 1.
[‡] Proceedings of the Munich Academy of Sciences, Math.-phys. Class 1891, **21**, p. 264.

and the condition for mechanical equilibrium of the atmosphere is

$$dp = \rho \, g \, dh. \tag{2}$$

Upon eliminating ρ from (1) and (2), we have

$$\frac{dp}{p} = \frac{M}{R} \cdot \frac{g}{t} \, dh. \tag{3}$$

(a) *Isothermal equilibrium.* For a general orientation, we consider isothermal equilibrium, that is, assume that t is constant. Then

$$p = p_0 \, exp \left(\frac{Mg \, h}{Rt} \right) \qquad \rho = \rho_0 \, exp \left(\frac{Mg \, h}{Rt} \right) \tag{4}$$

The gravity g on the sun is 27·7 times greater than on Earth, and the temperature (about 6000 °C) is about 20 times higher. Hence for a gas having the molecular weight of air, the spatial pressure distribution is approximately the same as for air on Earth. Upon closer analysis, the increase of pressure and density is found to be a factor of 10 every 14·7 km for a gas of the molecular weight of air (28·9), and every 212 km for hydrogen. Since 725 km on the sun correspond to an angle of 1 second of arc as seen from Earth, clearly the sun must appear sharply defined.

(b) *Adiabatic equilibrium.* If a gaseous mass expands adiabatically, the Poisson relations hold good

$$\frac{p}{p_0} = \left(\frac{\rho}{\rho_0} \right)^k = \left(\frac{t}{t_0} \right)^{\frac{k}{k-1}}, \tag{5}$$

where p_0 and ρ_0 are some connected initial values. The quantity k, the specific heat ratio, is 5/3 for a monatomic gas, 7/5 for a diatomic gas, 4/3 for a triatomic gas and it decreases to 1 for polyatomic gases. The equilibrium of an atmosphere is said to be adiabatic if the temperature is everywhere the one which an ascending and adiabatically cooling gaseous mass would take on, i.e. if equations (5) are satisfied throughout the entire atmosphere.

We then have from (3) after substitution of (5) and integration:

$$t - t_0 = \frac{k-1}{k} \ \frac{Mg}{R} (h - h_0).$$ (6)

The temperature varies linearly with altitude. The temperature gradient for the atmosphere of Earth is calculated to be 1 °C per 100 m, and for the sun this is 27·7 times greater than on Earth. The temperature increases thus by 1 °C every 3·63 m for air, and every 52 m for hydrogen. The atmosphere has a definite outer limit ($t = \rho = p = 0$). The depth below the outer limit of a layer at 6000 °C will, on the sun, be 22 km for air and 300 km for hydrogen.

(c) *Radiative equilibrium.* If one imagines the outer parts of the sun to form a continuous transition to an increasingly hotter and denser gaseous mass, no distinction can be drawn between radiating and absorbing layers and each layer has to be regarded as simultaneously radiative and absorbent. We know that a huge stream of energy originating from unknown sources in the interior of the sun passes through the solar atmosphere and streams into outer space. In the absence of mixing motions, what temperature must the individual layers of the solar atmosphere assume for such a stream of energy to be forwarded steadily without further variation of their own temperature?

Suppose that each altitudinal layer dh of the solar atmosphere absorbs the fraction adh of the radiation passing through it. If E is the emission of a black body at the temperature of this layer and one assumes that Kirchhoff's law applies, it follows that the layer will radiate the energy $E.a$dh in every direction.

Consider now the radiation energy A which at an arbitrary point is travelling outwards through the solar atmosphere, and the radiation energy B which (due to the radiation of the more outlying layers) is travelling inwards.

Let us first consider the energy B moving inwards. If one moves an infinitely thin layer dh inwards, then a fraction B.adh of the incoming energy B will be lost, on the other hand the quantity $a E$dh of the radiation of the layer dh itself in the same direction is added, therefore altogether we have

$$\frac{dB}{dh} = a(E - B). \tag{7}$$

Likewise for the outward radiation

$$\frac{dA}{dh} = -a(E - A). \tag{8}$$

Given that the absorptive capacity a is a function of the depth h, one can construct the "optical mass" of the atmosphere above the height h:

$$m = \int^{h} a\,dh. \tag{9}$$

We then have the differential equations

$$\frac{dB}{dm} = E - B, \quad \frac{dA}{dm} = A - E. \tag{10}$$

We are looking for a stationary state for the temperature distribution, and this state is conditioned by the requirement that each layer should receive as much energy as it emits, i.e. that

$$aA + aB = 2aE, \quad A + B = 2E.$$

If we introduce an auxiliary quantity γ to take this condition into account:

$$A = E + \gamma, \quad B = E - \gamma,$$

then by addition and subtraction the differential equations become

$$\frac{d\gamma}{dm} = 0, \quad \frac{dE}{dm} = \gamma.$$

Upon integration

$$\gamma = \text{const}, \quad E = E_0 + \gamma m,$$
$$A = E_0 + \gamma(1 + m), \quad B = E_0 + \gamma(m - 1).$$

The integration constants E_0 and γ have been so defined that no inward moving energy B is present at the outer limit of the atmosphere ($m = 0$), whilst the outward moving energy has the observed value A_0. Therefore, for $m = 0$,

$$B = 0, \quad A = A_0.$$

Hence

$$E=\frac{A_0}{2}(1+m), \quad A=\frac{A_0}{2}(2+m), \quad B=\frac{A_0}{2}m. \qquad (11)$$

The dependence of the radiation E on the optical mass above a given point can therefore be derived on the assumption of Kirchhoff's law only.

If one wishes to know the pressure and density distribution which prevails at radiative equilibrium, one fundamentally needs a detailed investigation which treats the radiation at the individual wavelengths. For a preliminary inquiry at this point it is sufficient to assume that the coefficient of absorption is independent of the colour and proportional to the density:

$$a=\rho k, \qquad (12)$$

where k is a constant. Then by definition

$$m = \int a \, dh = k \cdot \int \rho \, dh = \frac{k}{g} \cdot p. \qquad (13)$$

According to Stefan's law the black body radiation E is

$$E = c \cdot t^4 \ (c = \text{const}).$$

If one puts for the outwards moving energy A_0:

$$A_0 = c \cdot T^4,$$

then T is the quantity which is usually given as the (effective) temperature of the sum; approximately $T = 6000$ °C. For radiative equilibrium we then have, according to (11),

$$t^4 = \tfrac{1}{2}T^4\left[1 + \frac{k}{g}p\right]. \qquad (14)$$

Upon introducing the temperature τ at the outer limit of the atmosphere ($\tau = T/\sqrt[4]{2}$), one can write also

$$t^4 = \tau^4\left[1 + \frac{k}{g}p\right]. \qquad (14a)$$

Substituting this into (3), we obtain

$$\frac{M}{R}gh = \int^t \frac{4t^4 dt}{t^4 - \tau^4} = \tau\left[4\frac{t}{\tau} + \log\frac{\frac{t-1}{\tau}}{\frac{t+1}{\tau}} - 2 \arctan\frac{t}{\tau}\right] + \text{const.} \qquad (15)$$

This equation gives the temperature as a function of the depth. The associated density thus is

$$\rho = \frac{M}{R}\frac{p}{t} = \frac{Mg}{Rk}\frac{t^4 - \tau^4}{t\tau^4}. \tag{16}$$

The following table gives the values which are obtained from (11), (15) and (16) if the solar atmosphere is considered to consist of our air. The absorption coefficient for air is approximately $k = 0.6$, and the effective temperature $T = 6000$ °C, hence the temperature of the outer limit is $\tau = 5050$ °C. The depth h is calculated from a point where the temperature is $1\frac{1}{2}$ times the outer limit temperature. These define the basic data used in the calculation.

Depth h	$t.$	m	ρ
$-\infty$	5050°	0·000	0·00
-36.9 km	5060°	0·018	0·02
-19.1	5300°	0·215	0·51
0.0	7570°	4·06	6·8
$+12.0$	10,100°	15·0	18·7
$+55.7$	20,200°	255·0	159·4

In order to obtain the corresponding table for hydrogen, the depth h should be multiplied by 14·4, whilst the density, should on the one hand, be divided by 14·4 and, on the other hand, should be multiplied by a factor which takes into account how much the same mass of hydrogen is more transparent than air. The two columns t and m remain unchanged.

It will be seen that radiative equilibrium with increasing height above the sun approximates the isothermal equilibrium corresponding to the boundary temperature τ, and that it theoretically leads to an atmosphere of infinite extent as for isothermal equilibrium.

3. Stability of Radiative Equilibrium

A comparison of the temperature gradients at radiative and adiabatic equilibrium is of special interest. If, for instance, the temperature gradient is lower than at adiabatic equilibrium, an ascending air mass will enter layers which are warmer and thinner than itself on arrival, and so it will experience a downward pressure. Similarly a descending air mass experiences an upward pressure. An equilibrium at a temperature gradient below the adiabatic value is therefore stable, whilst conversely equilibrium is unstable at a temperature gradient above the adiabatic value.

For adiabatic equilibrium we have, in accordance with (6),

$$\frac{\mathrm{d}t}{\mathrm{d}h} = \frac{k-1}{k}\frac{Mg}{R}$$

and for radiative equilibrium, according to (15),

$$\frac{\mathrm{d}t}{\mathrm{d}h} = \tfrac{1}{4}\left(1 - \frac{4}{t^4}\right)Mg$$

The stability condition is therefore

$$1 - \frac{\tau^4}{t^4} < 4\,\frac{k-1}{k}$$

and this is always satisfied for $k > 4/3$.

Hence radiative equilibrium is everywhere stable so long as the gas forming the atmosphere is mon-, di-, or tri-atomic. For higher polyatomic cases instability would arise in deeper layers (of higher temperature t).

Thus it is reasonable to imagine an outer shell of the solar atmosphere in stable radiative equilibrium while perhaps in the interior there is a zone of ascending and descending currents which is close to adiabatic equilibrium and which at the same time takes care of the removal of energy.

4. Brightness Distribution over the Solar Disc

On our assumptions every vertical temperature distribution in the solar atmosphere corresponds to a brightness distribution over the solar disc.

We have previously considered the total energy A travelling out of the solar atmosphere without separating its individual components which move at different inclinations to the vertical, and we have denoted the coefficient of absorption for the total energy be a. This a is an average value of the absorption coefficients which hold for all possible inclinations.

We wish now to consider separately the radiation in a particular direction and we denote by $F(i)$ the radiation moving at an angle i to the vertical. Let the coefficient of absorption for the radiation which penetrates the atmosphere in the normal direction be denoted by α. Then $\alpha/\cos i$ is the coefficient of absorption for the radiation at the angle i. In full analogy with (8), the differential equation for F is therefore

$$\frac{\mathrm{d}F}{\mathrm{d}h} = -\frac{\alpha}{\cos i}(E-F), \tag{18}$$

or

$$\frac{\mathrm{d}F}{\mathrm{d}\mu} = -\frac{1}{\cos i}(E-F)$$

if the following notation is used

$$\mu = \int_{-\infty}^{h} \alpha \mathrm{d}h. \tag{19}$$

For the radiation emerging from the atmosphere we have by integration

$$F(i) = \int_{0}^{\infty} E e^{-\mu \, sec \, i} \, \mathrm{d}\mu \, sec \, i. \tag{20}$$

One can therefore evaluate $F(i)$ as soon as the temperature distribution along the vertical, and hence also E as a function of μ, is known.

However, μ is related to the optical mass m introduced previously. Consider the total radiation which is incident from below on a horizontal element of area $\mathrm{d}s$ within the atmosphere. It is given in terms of the integral over the radiation coming from all possible directions:

$$J = 2\pi ds \int_0^{\pi/2} di \sin i \cos i \, F(i).$$

The absorption which this radiation suffers within the layer dh will be

$$dJ = 2\pi ds \int_0^{\pi/2} di \sin i \cos i \, F(i) \frac{adh}{\cos i}$$

$$= 2\pi ds \, adh \int_0^{\pi/2} di \sin i \, F(i).$$

The coefficient of absorption a used previously for the total energy was defined by the relation

$$\frac{dJ}{J} = adh.$$

Upon comparison with the present formulae, we have

$$a = a \cdot \frac{\displaystyle\int_0^{\pi/2} di \sin i \, F(i)}{\displaystyle\int_0^{\pi/2} di \sin i \cos i \, F(i)} \cdot$$

If $F(i)$ is fairly constant for small inclinations i and only varies rapidly at values of i close to 90°—which is the case with the sun as follows from the experimental data discussed below—we obtain an approximation for a by regarding $F(i)$ as completely independent of i and integrating:

$$a = 2a. \tag{21}$$

We shall use this expression. From (9) and (19) we have

$$m = 2\mu$$

hence (20) becomes

$$F(i) = \int_0^\infty E \, exp\left(-\frac{m}{2} \sec i\right) \frac{dm}{2} \, \sec i.$$

Therefore $F(i)$ is known when E is given as a function of the optical mass m. However, the function $F(i)$ also yields directly the brightness distribution over the solar disc. The radiation which we observe on the solar disc at an apparent distance r from the centre of the disc, has obviously passed through the solar atmosphere at an angle i which is given by the equation

$$\sin i = \frac{r}{R}, \tag{23}$$

where R is the apparent solar radius. The combination of (22) and (23) associates the appropriate brightness F to each r.

The relation between the radiation distribution E in depth and the brightness distribution F on the surface is very clear if E is expanded as a power series in E

$$E = \rho_0 + m\rho_1 + m^2\rho_2 + \dots \,. \tag{24}$$

Then from (22) we have

$$F = \beta_0 + 2 \cdot 1! \, \beta_1 \cos i + 4 \cdot 2! \, \beta_2 \cos^2 i + \dots \tag{25}$$

If E can be represented as a sum of fractional powers of m:

$$E = \sum_n \beta \cdot m^{\gamma_n} \,, \tag{26}$$

then

$$F = \sum_n \Gamma(\gamma_n)\beta_n(2 \cos i)^{\gamma_n} \,, \tag{27}$$

where Γ is the Γ-function. Also here the transition from E to F presents no difficulty.

We want more particularly to consider the brightness distribution arising in radiative and adiabatic equilibrium.

For *radiative equilibrium,* according to (11), we have

$$E = \frac{A_0}{2}(1+m),$$

whence, according to (24) and (25),

$$F(i) = \frac{A_0}{2}(1+2\cos i),$$

or, taking the brightness at the centre of the solar disc $(i=0)$ as unity,

$$F(i) = \frac{1+2\cos i}{3}. \tag{28}$$

For *adiabatic equilibrium* we have equation (5):

$$\frac{t}{t_0} = \left(\frac{p}{p_0}\right)^{(k-1)/k}.$$

If, as above, the absorption is assumed to be equal for all colours and proportional to the density, we have the relation (13) between p and m:

$$m = \frac{k}{g}p.$$

We thus get for E

$$E = c \cdot t^4 = c_1 p^{4\frac{k-1}{k}} = c_2 m^{4\frac{k-1}{k}},$$

where c_1 and c_2 are new constants. The corresponding expression from F, according to (26) and (27), is

$$F(i) = c_2 \Gamma\left[4\frac{(k-1)}{k}\right](\cos i)^{4\frac{k-1}{k}},$$

or, again taking the central brightness as equal to unity,

$$F(i) = (\cos i)^{4 \cdot \frac{k-1}{k}}. \tag{29}$$

The formulae (28) *and* (29) *can be compared with observations.* Apart from spectro-photometric investigations for individual colour regions, there have been other relevant measurements carried out with thermopiles and bolometers which indicate how the total radiation added over all wavelengths is distributed over the solar disc. These measurements have been summarized by G. Müller in his "Photometrie der Gestirne" (*Photometry of the stars*), page 323, and his average values are reproduced below in the second column of the table. The theoretical values for radiative and adiabatic equilibrium calculated from equations (28) and (29) are shown in the two adjoining columns. For adiabatic equilibrium we put $k = 4/3$, which corresponds to a tri-atomic gas. Mon- or di-atomic gases would be even less in agreement, and physically it is improbable that the gases in the outer parts of the solar atmosphere should be more than tri-atomic.

$\dfrac{r}{R}$	Measured Value	Radiative Equilibrium	Adiabatic Equilibrium
0·0	1·00	1·00	1·00
0·2	0·99	0·99	0·98
0·4	0·97	0·95	0·92
0·6	0·92	0·87	0·80
0·7	0·87	0·81	0·71
0·8	0·81	0·73	0·60
0·9	0·70	0·63	0·44
0·96	0·59	0·52	0·28
0·98	0·49	0·47	0·20
1·00	(0·40)	0·33	0·00

We thus see that *radiative equilibrium represents the brightness distribution over the solar disc as well as can be expected on the simplified assumptions made in the present analysis, whilst we may expect that the adiabatic equilibrium would lead to quite a different picture of the solar disc.* Hence we have found a definite empirical justification for the introduction of radiative equilibrium.

E.S.P.—10*

27

The Earth and Sun as Magnets†

George Ellery Hale‡

In 1891 Prof. Arthur Schuster, speaking before the Royal Institution, asked a question which has been widely debated in recent years: "Is every large rotating body a magnet?" Since the days of Gilbert, who first recognized that the earth is a great magnet, many theories have been advanced to account for its magnetic properties. Biot, in 1805, ascribed them to a relatively short magnet near its center. Gauss, after an extended mathematical investigation, substituted a large number of small magnets, distributed in an irregular manner, for the single magnet of Biot. Grover suggested that terrestrial magnetism may be caused by electric currents, circulating around the earth and generated by the solar radiation. Soon after Rowland's demonstration in 1876, that a rotating electrically charged body produces a magnetic field, Ayrton and Perry attempted to apply this principle to the case of the earth. Rowland at once pointed out a mistake in their calculation, and showed that the high potential electric charge demanded by their theory could not possibly exist on the earth's surface. It remained for Schuster to suggest that a body made up of molecules which are neutral in the ordinary electrical or magnetic sense may nevertheless develop magnetic properties when rotated.

† *Smithsonian Report for 1913*, pp. 145–58. Address delivered at the semi-centennial of the National Academy of Sciences, at Washington, D.C., May 1913.

‡ The author had expected, before reprinting this address, to subject it to a thorough revision and to insert the results of recent observations, but he has been prevented by illness from doing so. (Aug. 24, 1914.)

We shall soon have occasion to examine the two hypotheses advanced in support of this view. While both are promising, it can not be said that either has been sufficiently developed to explain completely the principal phenomena of terrestrial magnetism. If we turn to experiment, we find that iron globes spun at great velocity in the laboratory fail to exhibit magnetic properties. But this can be accounted for on either hypothesis. What we need is a globe of great size, which has been rotating for centuries at high velocity. The sun, with a diameter 100 times that of the earth, may throw some light on the problem. Its high temperature wholly precludes the existence of permanent magnets, hence any magnetism it may exhibit must be due to motion. Its great mass and rapid linear velocity of rotation should produce a magnetic field much stronger than that of the earth. Finally, the presence in its atmosphere of glowing gases and the well-known effect of magnetism on light should enable us to explore its magnetic field even at the distance of the earth. The effects of ionization, probably small in the region of high pressure beneath the photosphere and marked in the solar atmosphere, must be determined and allowed for. But with this important limitation the sun may be used by the physicist for an experiment which can not be performed in the best equipped laboratory.

Schuster, in the lecture already cited, remarked:

"The form of the corona suggests a further hypothesis which, extravagant as it may appear at present, may yet prove to be true. Is the sun a magnet?"

Summing up the situation in April 1912, he repeated:

"The evidence (whether the sun is a magnet) rests entirely on the form of certain rays of the corona, which—assuming that they indicate the path of projecting particles—seem to be deflected as they would be in a magnetic field, but this evidence is not at all decisive."

There remained the possibility of an appeal to a conclusive test of magnetism—the characteristic changes it produces in light which originates in a magnetic field.

Before describing how this test has been applied, let us rapidly recapitulate some of the principal facts of terrestrial magnetism.

You see upon the screen the image of a steel sphere, which has been strongly magnetized. If iron filings are sprinkled over the glass plate that supports it, each minute particle becomes a magnet under the influence of the sphere. When the plate is tapped, to relieve the friction, the particles fall into place along the lines of force, revealing a characteristic pattern of great beauty. A small compass needle, moved about the sphere, always turns so as to point along the lines of force. At the magnetic poles it points toward the center of the sphere. Midway between them, at the equator, it is parallel to the diameter joining the poles.

As the earth is a magnet it should exhibit lines of force resembling those of the sphere. If the magnetic poles coincided with the poles of rotation, a freely suspended magnetic needle should point vertically downward at one pole, vertically upward at the other, and horizontally at the equator. A dip needle, used to map the lines of force of the earth, is shown on the screen. I have chosen for illustration an instrument designed for use at sea, on the nonmagnetic yacht *Carnegie*,† partly because the equipment used by Dr. Bauer in his extensive surveys represents the best now in use, and also because I wish to contrast the widely different means employed by the Carnegie Institution for the investigation of solar and terrestrial magnetic phenomena. The support of the dip needle is hung in gimbals, so that observations may be taken when the ship's deck is inclined. The smallest possible amount of metal enters into the construction of this vessel, and where its use could not be avoided, bronze was employed instead of iron or steel. She is thus admirably adapted for magnetic work, as is shown by the observations secured on voyages already totaling more than 100,000 miles. Her work is supplemented by that of land parties, bearing instruments to remote regions where magnetic observations have never before been made.

The dip needle clearly shows that the earth is a magnet, for it behaves in nearly the same way as the little needle used in our experiment with the magnetized sphere. But the magnetic poles

† Illustrated in article on The Earth's Magnetism, by Dr. Bauer, pp. 195–212 of this volume.

of the earth do not coincide with the geographical poles. The north magnetic pole, discovered by Ross and last visited by Amundsen in 1903, lies near Baffins Bay, in latitude 70° north, longitude 97° west. The position of the south magnetic pole, calculated from observations made in its vicinity by Capt. Scott, of glorious memory, in his expedition of 1901–1904, is 72° 50′ south latitude, 153° 45′ east longitude. Thus the two magnetic poles are not only displaced about 30° from the geographical poles; they do not even lie on the same diameter of the earth. Moreover, they are not fixed in position, but appear to be rotating about the geographical poles in a period of about 900 years. In addition to these pecularities, it must be added that the dip needle shows the existence of local magnetic poles, one of which has recently been found by Dr. Bauer's party at Treadwell Point, Alaska. At such a place the direction of the needle undergoes rapid change as it is moved about the local pole.

The dip needle, as we have seen, is free to move in a vertical plane. The compass needle moves in a horizontal plane. In general it tends to point toward the magnetic pole, and as this does not correspond with the geographical pole, there are not many places on the earth's surface where the needle indicates true north and south. Local pecularities, such as deposits of iron ore, also affect its direction very materially. Thus a variation chart, which indicates the deviation of the compass needle from geographical north, affords an excellent illustration of the irregularities of terrestrial magnetism. The necessity for frequent and accurate surveys of the earth's magnetic field is illustrated by the fact that the *Carnegie* has found errors of 5° or 6° in the variation charts of the Pacific and Indian Oceans.

In view of the earth's heterogeneous structure, which is sufficiently illustrated by its topographical features, marked deviations from the uniform magnetic properties of a magnetized steel sphere are not at all surprising. The phenomenon of the secular variation, or the rotation of the magnetic poles about the geographical poles, is one of the peculiarities toward the solution of which both theory and experiment should be directed.

Passing over other remarkable phenomena of terrestrial magnetism, we come to magnetic storms and auroras, which are almost certainly of solar origin.

Here is a photograph of part of the sun, as it appears in the telescope. Scattered over its surface are sun spots, which increase and decrease in number in a period of about 11.3 years. It is well known that a curve, showing the number of spots on the sun, is closely similar to a curve representing the variations of intensity of the earth's magnetism. The time of maximum sun spots corresponds with that of reduced intensity of the earth's magnetism, and the parallelism of the two curves is too close to be the result of accident. We may therefore conclude that there is some connection between the spotted area of the sun and the magnetic field of the earth.

We shall consider a little later the nature of sun spots, but for the present we may regard them simply as solar storms. When spots are numerous the entire sun is disturbed, and eruptive phenomena, far transcending our most violent volcanic outbursts, are frequently visible. In the atmosphere of the sun, gaseous prominences rise to great heights. This one, reaching an elevation of 85,000 miles, is of the quiescent type, which changes gradually in form and is abundantly found at all phases of the sun's activity. But such eruptions as the one of March 25, 1895, photographed with the spectroheliograph of the Kenwood Observatory, are clearly of an explosive nature. As these photographs show, it shot upward through a distance of 146,000 miles in 24 minutes, after which it faded away.

When great and rapidly changing spots, usually accompanied by eruptive prominences, are observed on the sun, brilliant displays of the aurora and violent magnetic storms are often reported. The magnetic needle, which would record a smooth straight line on the photographic film if it were at rest, trembles and vibrates, drawing a broken and irregular curve. Simultaneously, the aurora flashes and pulsates, sometimes lighting up the northern sky with the most brilliant display of red and green discharges.

Birkeland and Störmer have worked out a theory which accounts in a very satisfactory way for these phenomena. They suppose that electrified particles, shot out from the sun with great velocity, are drawn in toward the earth's magnetic poles along the lines of force. Striking the rarified gases of the upper atmosphere, they illuminate them, just as the electric discharge lights up a vacuum tube. There is reason to believe that the highest part of the earth's atmosphere consists of rarified hydrogen, while nitrogen predominates at a lower level. Some of the electrons from the sun are absorbed in the hydrogen, above a height of 60 miles. Others reach the lower-lying nitrogen, and descend to levels from 30 to 40 miles above the earth's surface. Certain still more penetrating rays sometimes reach an altitude of 25 miles, the lowest hitherto found for the aurora. The passage through the atmosphere of the electrons which cause the aurora also gives rise to the irregular disturbances of the magnetic needle observed during magnetic storms.

The outflow of electrons from the sun never ceases, if we may reason from the fact that the night sky is at all times feebly illuminated by the characteristic light of the aurora. But when sun spots are numerous, the discharge of electrons is most violent, thus explaining the frequency of brilliant auroras and intense magnetic storms during sun-spot maxima. It should be remarked that the discharge of electrons does not necessarily occur from the spots themselves, but rather from the eruptive regions surrounding them.

Our acquaintance with vacuum-tube discharges dates from an early period, but accurate knowledge of these phenomena may be said to begin with the work of Sir William Crookes in 1876. A glass tube, fitted with electrodes, and filled with any gas, is exhausted with a suitable pump until the pressure within it is very low. When a high-voltage discharge is passed through the tube, a stream of negatively charged particles is shot out from the cathode, or negative pole, with great velocity. These electrons, bombarding the molecules of the gas within the tube, produce a brilliant illumination, the character of which depends upon the nature of the gas. The rare hydrogen gas in the upper atmosphere of the earth, when bombarded by electrons from the sun, glows like the hydrogen in this tube. Nitrogen, which is characteristic of a lower level, shines with the light which can be duplicated here.

But it may be remarked that this explanation of the aurora is only hypothetical, in the absence of direct evidence of the emission of electrons by the sun. However, we do know that hot bodies emit electrons. Here is a carbon filament in an exhausted bulb. When heated white hot a stream of electrons passes off. Falling upon this electrode the electrons discharge the electroscope with which it is connected. Everyone who has to discard old incandescent lamps is familiar with the result of this outflow. The blackening of the bulbs is due to finely divided carbon carried away by the electrons and deposited upon the glass.

Now, we know that great quantities of carbon in a vaporous state exist in the sun and that many other substances also present there emit electrons in the same way. Hence we may infer that electrons are abundant in the solar atmosphere.

The temperature of the sun is between 6000° and 7000° C, twice as high as we can obtain by artificial means. Under solar conditions, the velocity of the electrons emitted in regions where the pressure is not too great may be sufficient to carry them to the earth. Arrhenius holds that the electrons attach themselves to molecules or groups of molecules and are then driven to the earth by light pressure.

In certain regions of the sun we have strong evidence of the existence of free electrons. This leads us to the question of solar magnetism and suggests a comparison of the very different conditions in the sun and earth. Much alike in chemical composition, these bodies differ principally in size, in density, and in temperature. The diameter of the sun is more than 100 times that of the earth, while its density is only one-quarter as great. But the most striking point of difference is the high temperature of the sun, which is much more than sufficient to vaporize all known substances. This means that no permanent magnetism, such as is exhibited by a steel magnet or a lodestone, can exist in the sun. For if we bring this steel magnet to a red heat it loses its magnetism and drops the iron bar which it previously supported. Hence, while some theories attribute terrestrial magnetism to the presence within the earth of permanent magnets, no such theory can apply to the sun. If magnetic phenomena are to be found there they must result from other causes.

The familiar case of the helix illustrates how a magnetic field is produced by an electric current flowing through a coil of wire. But according to the modern theory, an electric current is a stream of electrons. Thus a stream of electrons in the sun should give rise to a magnetic field. If the electrons were whirled in a powerful vortex, resembling our tornadoes or waterspouts, the analogy with the wire helix would be exact, and the magnetic field might be sufficiently intense to be detected by spectroscopic observations.

A sun spot, as seen with a telescope or photographed in the ordinary way, does not appear to be a vortex. If we examine the solar atmosphere above and about the spots, we find extensive clouds of luminous calcium vapor, invisible to the eye, but easily photographed with the spectroheliograph by admitting no light to the sensitive plate except that radiated by calcium vapor. These calcium flocculi, like the cumulus clouds of the earth's atmosphere, exhibit no well-defined linear structure. But if we photograph the sun with the red light of hydrogen, we find a very different condition of affairs. In this higher region of the solar atmosphere, first photographed on Mount Wilson in 1908, cyclonic whirls, centering in sun spots, are clearly shown.

The idea that sun spots may be solar tornadoes, which was strongly suggested by such photographs, soon received striking confirmation. A great cloud of hydrogen, which had hung for several days on the edge of one of these vortex structures, was suddenly swept into the spot at a velocity of about 60 miles per second. More recently Slocum has photographed at the Yerkes Observatory a prominence at the edge of the sun, flowing into a spot with a somewhat lower velocity.

Thus we were led to the hypothesis that sun spots are closely analogous to tornadoes or waterspouts in the earth's atmosphere. If this were true, electrons caught and whirled in the spot vortex should produce a magnetic field. Fortunately, this could be put to a conclusive test through the well-known influence of magnetism on light discovered by Zeeman in 1896.

In Zeeman's experiment a flame containing sodium vapor was placed between the poles of a powerful electromagnet. The two yellow sodium lines, observed with a spectroscope of high dispersion, were seen to widen the instant a magnetic field was produced by passing a current through the coils of the magnet. It was subsequently found that most of the lines of the spectrum, which are single under ordinary conditions, are split into three components when the radiating source is in a sufficiently intense magnetic field. This is the case when the observation is made at right angles to the lines of force. When looking along the lines of force, the central line of such a triplet disappears, and the light of the two side components is found to be circularly polarized in opposite directions. With suitable polarizing apparatus, either component of such a line can be cut off at will, leaving the other unchanged. Furthermore, a double line having these characteristic properties can be produced only by a magnetic field. Thus it becomes a simple matter to detect a magnetic field at any distance by observing its effect on light emitted within the field. If a sun spot is an electric vortex, and the observer is supposed to look along the axis of the whirling vapor, which would correspond with the direction of the lines of force, he should find the spectrum lines double, and be able to cut off either component with the polarizing attachment of his spectroscope.

I applied this test to sun spots on Mount Wilson in June, 1908, with the 60-foot tower telescope, and at once found all of the characteristic features of the Zeeman effect. Most of the lines of the sun-spot spectrum are merely widened by the magnetic field, but others are split into separate components, which can be cut off at will by the observer. Moreover, the opportune formation of two large spots, which appeared on the spectroheliograph plates to be rotating in opposite directions, permitted a still more exacting experiment to be tried. In the laboratory, where the polarizing apparatus is so adjusted as to transmit one component of a line doubled by a magnetic field, this disappears and is replaced by the other component when the direction of the current is reversed. In other words, one component is visible alone when the observer looks toward the north pole of the magnet, while the other appears alone when he looks toward the south pole. If electrons of the same kind are rotating in opposite directions in two sun-spot vortexes, the observer should be looking toward a north pole in one spot and toward a south pole in the other. Hence the opposite components of a magnetic double line should appear in two such spots. As our photographs show, the result of the test was in harmony with my anticipation.

I may not pause to describe the later developments of this investigation, though two or three points must be mentioned. The intensity of the magnetic field in sun spots is sometimes as high as 4500 gausses, or 9000 times the intensity of the earth's field. In passing upward from the sun's surface the magnetic intensity decreases very rapidly—so rapidly, in fact as to suggest the existence of an opposing field. It is probable that the vortex which produces the observed field is not the one that appears on our photograph, but lies at a lower level. In fact, the vortex structure shown on spectroheliograph plates may represent the effect rather than the cause of the sun-spot field. We may have, as Brester and Deslandres suggest, a condition analogous to that illustrated in the aurora: Electrons, falling in the solar atmosphere, move along the lines of force of the magnetic field into spots. In this way we may perhaps account for the structure surrounding pairs of spots, of opposite polarity, which constitute the typical sun-spot group. The resemblance of the structure near these two bipolar groups to the lines of force about a bar magnet is very striking, especially when the disturbed condition of the solar atmosphere, which tends to mask the effect, is borne in mind. It is not unlikely that the bipolar group is due to a single vortex, of the horseshoe type, such as we may see in water after every sweep of an oar.

We thus have abundant evidence of the existence on the sun of local magnetic fields of great intensity—fields so extensive that the earth is small in comparison with many of them. But how may we account for the copious supply of electrons needed to generate the powerful currents required in such enormous electromagnets? Neutral molecules, postulated in theories of the earth's field, will not suffice. A marked preponderance of electrons of one sign is clearly indicated.

An interesting experiment, due to Harker, will help us here. Imagine a pair of carbon rods insulated within a furnace heated to a temperature of two or three thousand degrees. The outer ends of the rods projecting from the furnace are connected to a galvanometer. Harker found that when one of the carbon terminals within the furnace was cooler than the other a stream of negative electrons flowed toward it from the hotter electrode. Even at atmospheric pressure currents of several amperes were produced in this way.†

Our spectroscopic investigations, interpreted by laboratory experiments, are in harmony with those of Fowler in proving that sun spots are comparatively cool regions in the solar atmosphere. They are hot enough, it is true, to volatilize such refractory elements as titanium, but cool enough to permit the formation of certain compounds not found elsewhere in the sun. Hence, from Harker's experiment, we may expect a flow of negative electrons toward spots. These, caught and whirled in the vortex, would easily account for the observed magnetic fields.

The conditions existing in sun spots are thus without any close parallel among the natural phenomena of the earth. The sun-spot vortex is not unlike a terrestrial tornado, on a vast scale, but if the whirl of ions in a tornado produces a magnetic field, it is too feeble to be readily detected. Thus, while we have demonstrated the existence of solar magnetism, it is confined to limited areas. We must look further if we would throw new light on the theory of the magnetic properties of rotating bodies.

This leads us to the question with which we started: Is the sun a magnet, like the earth? The structure of the corona, as revealed at total eclipses, points strongly in this direction. Remembering the lines of force of our magnetized steel sphere, we can not fail to be struck by their close resemblance to the polar streamers in

† King has recently found that the current decreases very rapidly as the pressure increases, but is still appreciable at a pressure of 20 atm.

these beautiful photographs of the corona taken by Lick Observatory eclipse parties, for which I am indebted to Prof. Campbell. Bigelow, in 1889, investigated this coronal structure and showed that it is very similar to the lines of force of a spherical magnet. Störmer, guided by his own researches on the aurora, has calculated the trajectories of electrons moving out from the sun under the influence of a general magnetic field and compared these trajectories with the coronal streamers. The resemblance is apparently too close to be the result of chance. Finally, Deslandres has investigated the forms and motion of solar prominences, which he finds to behave as they would in a magnetic field of intensity about one-millionth that of the earth. We may thus infer the existence of a general solar magnetic field. But since the sign of the charge of the outflowing electrons is not certainly known, we can not determine the polarity of the sun in this way. Furthermore, our present uncertainty as to the proportion at different levels of positive and negative electrons and of the perturbations due to currents in the solar atmosphere must delay the most effective appication of these methods, though they promise much future knowledge of the magnetic field at high levels in the solar atmosphere.

Of the field at low levels, however, they may tell us little or nothing, for the distribution of the electrons may easily be such as to give rise to a field caused by the rotation of the solar atmosphere, which may oppose in sign the field due to the rotation of the body of the sun. To detect this latter field, the magnetic field of the sun as distinguished from that of the sun's atmosphere, we must resort to the method employed in the case of sun spots— the study of the Zeeman effect. If this is successful it will not only show beyond doubt whether the sun is a magnet; it will also permit the polarity of the sun to be compared with that of the earth, give a measure of the strength of the field at different latitudes and indicate the sign of the charge that a rotating sphere must possess if it is to produce a similar field.

I first endeavored to apply this test with the 60-foot tower telescope in 1908, but the results were too uncertain to command confidence.

Thanks to additional appropriations from the Carnegie Institute of Washington, a new and powerful instrument was available on Mount Wilson for a continuation of the investigation in January, 1912. The new tower telescope has a focal length of 150 feet. To prevent vibration in the wind, the coelostat, second mirror, and object glass are carried by a skeleton tower, each vertical and diagonal member of which is inclosed within the corresponding member of an outer skeleton tower, which also carries a dome to shield the instruments from the weather. In the photograph we see only the hollow members of the outer tower. But within each of them, well separated from possible contact, a sectional view would show the similar but more slender members of the tower that support the instruments. The plan has proved to be successful, permitting observations demanding the greatest steadiness of the solar image to be made.

The arrangements are similar to those of the 60-foot tower. The solar image, $16\frac{1}{2}$ inches in diameter, falls on the slit of a spectrograph in the observation house at the ground level. The spectrograph, of 75 feet focal length, enjoys the advantage of great stability and constancy of temperature in its subterranean vault beneath the tower. In the third order spectrum, used for this investigation, the D lines of the solar spectrum are 29 millimeters apart. The resolving power of the excellent Michelson grating is sufficient to show 75 lines of the iodine absorption spectrum in this space between the D's. Thus the instruments are well suited for the exacting requirements of a difficult investigation. For it must be borne in mind that the problem is very different from that of detecting the magnetic fields in sun spots, where the separation of the lines is from 50 to 100 times as great as we may expect to find here. Thus the sun's general field can produce no actual separation of the lines. But it may cause a very slight widening, which should appear as a displacement when suitable polarizing apparatus is used. This is so arranged as to divide the spectrum longitudinally into narrow strips. The component toward the red end of the spectrum of a line widened by magnetism should appear in one strip, the other component in the next strip. Hence, if the sun has a magnetic field of sufficient strength, the line should have a dentated appearance. The small relative displacements of the lines on successive strips, when measured under a microscope, should give the strength of the magnetic field.

The above remarks apply strictly to the case when the observer is looking directly along the lines of force. At other angles neither component is completely cut off, and the magnitude of the displacement will then depend upon two things: The strength of the magnetic field and the angle between the line of sight and the lines of force. Assuming that the lines of force of the sun correspond with those of a magnetized sphere, and also that the magnetic poles coincide with the poles of rotation, it is possible to calculate what the relative displacement should be at different solar latitudes. These theoretical displacements are shown graphically by the sine curve on the screen.

We see from the curve that the greatest displacements should be found at 45° north and south latitude, and that from these points they should decrease toward zero at the equator and the poles. Furthermore, the curve shows that we may apply the same crucial test used in the case of sun spots; the direction of the displacements, toward red or violet, should be reversed in the northern and southern hemispheres.

I shall not trouble you with the details of the hundreds of photographs and the thousands of measures which have been made by my colleagues and myself during the past year. In view of the diffuse character of the solar lines under such high dispersion and the exceedingly small displacements observed, the results must be given with some reserve, though they appear to leave no doubt as to the reality of the effect. Observations in the second order spectrum failed to give satisfactory indications of the field. But with the higher dispersion of the third order 11 independent determinations, made with every possible precaution to eliminate bias, show opposite displacements in the northern and southern hemispheres decreasing in magnitude from about 45° north and south latitude to the equator. Three of these determinations were pushed as close to the poles as conditions would permit, and the observed displacements may be compared with the theoretical curve. In view of the very small magnitude of the displacements, which never surpass 0.002 Ångströms, the agreement is quite as satisfactory as one could expect for a first approximation.

The full details of the investigation are given in a paper recently published.† The reader will find an account of the precautions taken to eliminate error, and, I trust, no tendency to underestimate the possible adverse bearing of certain negative results. It must remain for the future to confirm or to overthrow the apparently strong evidence in favour of the existence of a true Zeeman effect, due to the general magnetic field of the sun. If this evidence can be accepted, we may draw certain conclusions of present interest.

Taking the measures at their face value, they indicate that the north magnetic pole of the sun lies at or near the north pole of rotation, while the south magnetic pole lies at or near the south pole of rotation. In other words, if a compass needle could withstand the solar temperature, it would point approximately as it does on the earth, since the polarity of the two bodies appears to be the same. Thus, since the earth and sun rotate in the same direction, a negative charge distributed through their mass would account in each case for the observed magnetic polarity.

As for the strength of the sun's field, only three preliminary determinations have yet been made, with as many different lines. Disregarding the systematic error of measurement, which is still very uncertain, these indicate that the field strength at the sun's poles is of the order of 50 gauss (about 80 times that of the earth).

Schuster, assuming the magnetic fields of the earth and sun to be due to their rotation, found that the strength of the sun's field should be 440 times that of the earth, or 264 gauss. This was on the supposition that the field strength of a rotating body is proportional to the product of the radius and the maximum linear velocity of rotation, but neglected the density. Before inquiring why the observed and theoretical values differ, we may glance at

† *Contributions from the Mount Wilson Solar Observatory*, No. 71.

the two most promising hypotheses that have been advanced in support of the view that every large rotating body is a magnet.

On account of their greater mass, the positive electrons of the neutral molecules within the earth may perhaps be more powerfully attracted by gravitation than the negative electrons. In this case the negative charge of each molecule should be a little farther from the centre of the earth than the positive charge. The average linear velocity of the negative charge would thus be a little greater, and the magnetizing effect due to its motion would slightly exceed that due to the motion of the positive charge. By assuming a separation of the charges equal to about four-tenths the radius of a molecule (Bauer), the symmetrical part of the earth's magnetic field could be accounted for as the result of the axial rotation.

This theory, first suggested by Thomson, has been developed by Sutherland, Schuster, and Bauer. But as yet it has yielded no explanation of the secular variation of the earth's magnetism, and the merits of other theories must not be overlooked.

Chief among these is the theory that rests on the very probable assumption that every molecule is a magnet. If the magnetism is accounted for as the effect of the rapid revolution of electrons within the molecule, a gyrostatic action might be anticipated. That is, each molecule would tend to set itself with its axis parallel to the axis of the earth, just as the gyrostatic compass, now coming into use at sea, tends to point to the geographical pole. The host of molecular magnets, all acting together, might account for the earth's magnetic field.

This theory, in its turn, is not free from obvious points of weakness, though they may disappear as the result of more extended investigation. Its chief advantage lies in the possibility that it may explain the secular variation of the earth's magnetism by a precessional motion of the magnetic molecules.

On either hypothesis, it is assumed, in the absence of knowledge to the contrary, that every molecule contributes to the production of the magnetic field. Thus the density of the rotating body may prove to be a factor. Perhaps the change of density from the surface to the centre of the sun must also be taken into account. But the observational results already obtained suggest that the phenomena of ionization in the solar atmosphere may turn out to be the predominant influence.

The lines which show the Zeeman effect originate at a comparatively low level in the solar atmosphere. Preliminary measures indicate that certain lines of titanium, which are widely separated by a magnetic field in the laboratory, are not appreciably affected in the sun. As these lines represent a somewhat higher level, it is probable that the strength of the sun's field decreases very rapidly in passing upward from the surface of the photosphere—a conclusion in harmony with results obtained from the study of the corona and prominences. Thus it may be found that the distribution of the electrons is such as to give rise to the observed field or to produce a field opposing that caused by the rotation of the body of the sun. It is evident that speculation along these lines may advantageously await the accumulation of observations covering a wide range of level. Beneath the photosphere, where the pressure is high, we may conclude from recent electric furnace experiments by King that free electrons, though relatively few, may nevertheless play some part in the production of the general magnetic field.

In this survey of magnetic phenomena we have kept constantly in mind the hypothesis that the magnetism of the earth is due to its rotation. Permanent magnets, formerly supposed to account for the earth's magnetic field, could not exist at the high temperature of the sun. Displays of the aurora, usually accompanied by magnetic storms, are plausibly attributed to electrons reaching the earth from the sun, and illuminating the rare gases of the upper atmosphere just as they affect those in a vacuum tube. Definite proof of the existence of free electrons in the sun is afforded by the discovery of powerful local magnetic fields in sun spots, where the magnetic intensity is sometimes as great as nine thousand times that of the earth's field. These local fields probably result from the rapid revolution in a vortex of negative electrons, flowing toward the cooler spot from the hotter region outside. The same method of observation now indicates that the whole sun is a magnet, of the same polarity as the earth. Because of the high solar temperature, this magnetism may be ascribed to the sun's axial rotation.† It is not improbable that the earth's magnetism also results from its rotation, and that other rotating celestial bodies, such as stars and nebulae, may ultimately be found to possess magnetic properties. Thus, while the presence of free electrons in the sun prevents our acceptance of the evidence as a proof that every large rotating body is a magnet, the results of the investigation are not opposed to this hypothesis, which may be tested further by the study of other stars of known diameter and velocity of rotation.

† The alternative hypothesis, that the sun's magnetism is due to the combined effect of numberless local magnetic fields, caused by electric vortices in the solar "pores", though at first sight improbable, deserves further consideration.

Index